GETTING THE WILDERNESS IN YOU

BY WILLIAM M. SANDERSON II

Copyright © 2012 by William M. Sanderson II

Getting the Wilderness in You
by William M. Sanderson II

Printed in the United States of America

ISBN 9781622302536

All rights reserved solely by the author. The author guarantees all contents are original and do not infringe upon the legal rights of any other person or work. No part of this book may be reproduced in any form without the permission of the author. The views expressed in this book are not necessarily those of the publisher.

www.xulonpress.com

Getting the WILDERNESS in You

*"The way of the canoe
is the way of the wilderness
and of a freedom almost forgotten.
It is an antidote to insecurity,
the open door to
waterways of ages past,
and a way of life with profound
and abiding satisfactions.*

*When a man is part of his canoe,
he is part of all that canoes have ever known."*

Sigurd Olson,
Minnesota author and conservationist

The original brochure for Way of the Wilderness Canoe Outfitters

WHAT PEOPLE ARE SAYING

"This is a wonderful little book that reminds the reader that the beauty of both nature and friendship are treasures to embrace."
Barbara Hoover, Forensic Scientist

"Bill Sanderson's story of his experiences with Rolf and the Northwoods is an enjoyable read, full of humor, adventure, and great camping tips. More importantly, it demonstrates how the wilderness became a rite of passage from boy to young man. This is one of the few remaining places of solitude left on earth that gives us good reason to revisit."
Laura and Tom Lagana, Co-authors of *Chicken Soup for the Volunteer's Soul* and *Serving Productive Time*

"A captivating American tale that captures the joy of discovery through the enthusiastic eyes of a young boy. The details of the stories are so vivid that it carries you into the wilderness along with all the characters. There is an overall nostalgic hue ingrained in this book that makes you long for your youth, while appreciating your journey to adulthood."
Frances Kettering, Operations Manager, Mutual Health Services

"This book is extremely readable and makes you wish you had plans for a summer camping trip to the Northwoods!"
Nancy R. Norris, Accountant for the Salvation Army Northeast Ohio Divisional Headquarters

"A Great American story, not only with camping tips galore, but also with warm stories and adventures in the wonderful history of Northern Minnesota where we find what America is all about through the people who make us who we are."
Michael Stratton, Teacher, sailor, outdoorsman

"This is a great American story about the woodlands and waterways of Northern Minnesota chocked full of warm, humorous adventure stories and camping tips galore and experiences with people who make us who we are as we grow."
Cheryl Blanton, Legal Secretary, full-time camper

ACKNOWLEDGEMENTS

First, I wish to thank my wife Lynn for encouraging me in this book-writing endeavor. She has not only been patient as the writing and rewriting process unfolded, but was a tremendous help with the final details of publication and marketing.

I most certainly wish to thank Rolf and Gail Skrien for introducing me to this wonderful part of our earth. Their warm and gracious hospitality to me and to my family and friends was always present over the years. Letting me be part of their family for the summer I worked for them has a special place in my heart, and truly was part of the inspiration for the book. Sandra, Stewart, Stanton, Susan, Sally, and their families also helped review this work and provided valuable feedback for the stories.

Uncle Bud and Aunt Nancy, who took us to this place and let us take canoe trips as young boys, provided the opportunity for all of these experiences to occur. I know they realized how much these trips meant. My cousins, Ed, Dick, Bob, and Jim are like brothers to me, partially because we became men because of our experiences in the wilderness.

My friend, Mike Stratton provided valuable suggestions for the style and format used in the book, did extensive editing, and stayed with me for the years of writing this book.

Nancy Norris was my initial editor and spent hours reviewing and scrubbing the first manuscript, and Melody Detwiler and Ruth Johnson worked through a later version.

My son, Rob Sanderson provided the image for the cover of the book, and he and Dick Sanderson provided many of the other images used in the book as well. They both have an eye for pictures that is unique and special.

My daughter, Cara Sanderson spent hours working on the artwork and scrutinizing the book for design issues. She claims she is a one-person army of designers, and I agree.

Nick Reinfeld was a great encouragement and help during the initial publishing process. He also worked closely with the website design and marketing plans.

Thank you to the many attendees on the canoe trips where I read the manuscript aloud around the campfire. These opportunities helped shape the final product.

There are many others who help with projects we undertake like this one in particular, and not being mentioned in no way diminishes the importance of each special contribution. Thanks to all of you who have been a part of this project in any way.

Last, a special thank you to Penny Fox-Stewart, my senior editor, who patiently helped me polish the final version and make it ready for publication. Many phone calls and meetings were necessary to manage a project of this kind. Her commitment to every detail and oversight of the timeline helped bring the manuscript to a standard worthy of publication. I recommend her to anyone looking for a seasoned professional editor.

TABLE OF CONTENTS

WHAT PEOPLE ARE SAYING VII
ACKNOWLEDGEMENTS VIII
FORWARD 3
CHAPTER 1 - THE CALL 8
    ON THE WAY 12
    ED'S STORY 17
    HEADING NORTH 19
    UP THE TRAIL 24
CHAPTER 2 - MEETING ROLF AND GAIL 27
    TO CLARK ISLAND 38
    PREPARING FOR THE TRIP 42
CHAPTER 3 - THE FIRST TRIP 57
CHAPTER 4 - ROLF COMES TO VISIT 69
    ROLF TELLS HIS STORY 70
    HEADING HOME 87
CHAPTER 5 - BACK TO THE WILDERNESS 93
    THE EIGHT-DAY TRIP 96
CHAPTER 6 - SUMMER WITH ROLF AND GAIL 110
    LEARNING MORE 113
CHAPTER 7 - MORE ABOUT THE GUNFLINT TRAIL 125
    HOW GAIL AND ROLF MET 129
    THE VOYAGEURS COME THROUGH 133
    THE FINAL DAYS OF SUMMER 140
CHAPTER 8 - FAST FORWARD 149
CHAPTER 9 - ONE MORE TIME 164
EPILOGUE 178
BIBLIOGRAPHY 183
APPENDIX 185
ORDERING INFORMATION 202
ABOUT THE AUTHOR 203

FORWARD

All of us have things happen to us that are life changing. Unfortunately, sometimes these events are unforeseen and tragic. In my case, both of my parents died tragically within seven months of each other, and my sister and I were orphans.

Dealing with the death and absence of my parents has been a lifelong process, and I miss not having them around. One thing I have noticed, though, is that the provision and grace of my Creator has always been sufficient and timely to help me navigate through life's difficult journeys. I have never been alone. Family, however defined, is at the heart of God's plan for our well-being, even when part of that family is gone. People we encounter along our journey, whether they are friends, colleagues, or casual acquaintances, can make an enormous difference in our lives if we are open to it.

Some of the major events in my life that contributed to my transition from boy to young man include the camping and canoe trips with my Uncle Bud and Aunt Nancy and their boys to the Boundary Waters Canoe Area Wilderness located on the U.S.-Canadian border in Minnesota, also known as the BWCAW, or simply, the Boundary Waters. The pristine beauty of this area is still relatively unchanged although the lodges and other businesses have modern conveniences for those visitors who do not want to experience so much raw wilderness. The wilderness, however, offers none of these conveniences.

When I returned to the Boundary Waters nearly thirty years after my earlier visits in the 1960s, I found it much the same as I remembered it. I was surprised to see many of the people and businesses still there from when I was younger. The present confirmed the existence of the past.

A friend of mine had a similar experience when he returned to the one-room school he attended as a boy in

Iowa. During a trip there, he decided to go by the place where the school was to see if anything still existed after forty-five years. Although the building was overgrown and boarded, he found an old textbook in one of the desks that had his name written inside the front cover. It was validation that he had been there and had passed that way before.

Just as important as the events and places in our past are the people who left lasting impressions on us. I was surprised to see so many of the same folks I remember still living and working there.

My first goal in writing this book is to create something to read and enjoy because of the true stories of my journey from boy to young man. As I get older, I often hear my children, grandchildren, family members, and even friends comment about hearing stories "a million times!" There are often quiet, subtle, and almost unknown contributions from people or experiences that have forever touched our personality and way of being. Without these stories, our lives lose a dimension of richness and meaning. It is important that we pass our stories and experiences on to others, because these accounts help us remember our own history, and the people, places, and situations that have helped us become who we are and help us transition from boys to men and girls to women. In some sense, these stories are like bookmarks in our lives, holding a special place where we return every now and again for reference or just for the fun of remembering.

My second goal is to introduce you to a wonderful man named Rolf Skrien. He and his wife Gail are the inspiration for this book. I met Rolf during my first time at the Boundary Waters, and he was my canoe outfitter, employer, and mentor. He is one of the people in my life who had a deep and lasting influence in my journey from boy to man.

Still a very unassuming man at ninety-one years old, Rolf has touched many lives because of his attitude toward life and work. He started his canoe outfitting business in 1956, and kept it for twenty years. He remembers the people and country as clearly as ever and still gets letters and visits from former customers. He still tells stories of the area and his experiences, and helps route my annual canoe trips. Because of his love of the wilderness and his commitment to sharing what he knows about it and instilling that love in others, he has successfully passed the baton to fellow travelers. It is my goal to pass the baton to others as well, just as I have shared my love of canoeing with my family and others.

Although parts of their stories are in several publications[1], it seems only fitting that someone write about Rolf and Gail in one book because of the way they have touched the lives of so many people over the years. Rolf is always thrilled to hear the stories of canoe trips from his former customers. He once said that he felt like a teacher contributing to the enjoyment of nature and helping people learn about the great outdoors. I want to pass along the same story, but in a way that lets the reader see the influence Rolf and Gail have had on my life.

I have consulted with many individuals in writing this book to ensure the accuracy of events and timeline so it is easy to see the connection and influence people like Rolf and Gail have had with their customers. The book is written in a first-person narrative format because tape recordings of original conversations from decades ago are not available and memories are not always reliable. I painstakingly interviewed the sources of these stories to obtain the best possible recollections, and spent hours interviewing and tape recording Rolf to gather a compilation of his knowledge and recollections of this area. Capturing his experiences in this way has been an experience in itself.

---

[1]See Bibliography, in particular, books by Hunt, Webster, Muus, and The Women of the Gunflint Trail.

Along with my own personal diaries, logs, additional research, and reading, most of the historical facts are accurate. I inserted certain details in places in the story where they most likely would have occurred or provided information to the reader to help develop an understanding. I have also included actual taped interviews with Rolf, and his stories are interwoven in the book at specific places to add his perspective.

My challenge to you as you read this story is to look for opportunities to share your story with others. What events have shaped the course of your life? What places have special meaning for you? Who are the people whose lives have made an indelible impression on yours?

We may never realize how much of a difference we make in someone's life. We may never see the healing that came through our hand to someone in pain from the hardships in their life. But we are never alone.

Enjoy our time together as we explore the Minnesota wilderness and remember to pass the baton.

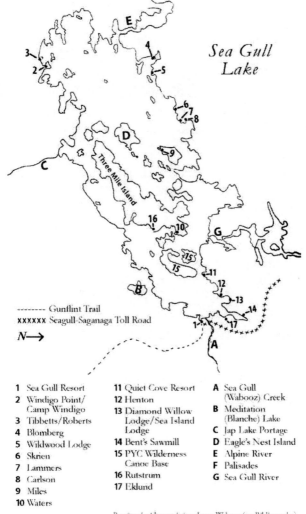

Map of Sea Gull Lake showing key locations

CHAPTER 1 - THE CALL

I remember being in the living room when the phone rang. It was a gloomy day in Cleveland, Ohio, and it was 1966. The first signs of spring were evident as trees showed their buds and the daffodils had poked up from the ground. April was such a tease with these signs of spring all around, but the chilly, gray fingers of winter would not loosen their grip easily.

My sister, Diane, answered the phone. She was my older half sister, and became my guardian after Mom and Dad died. My mother had been married before and had two daughters, Diane and Linda. Diane was sixteen years older than I was and had just started a family with her husband, Don when she took my younger sister, Cindy and me to live with them after we were orphaned. Diane and I always got along well. She was supportive of most everything I did and I looked at her as a combination mother/sister, but mainly as a sister. She trusted me to make good choices, and I really appreciated that.

There was a brief pause, then the voice of recognition.

"Nancy! It's great to hear your voice!" Diane said. "We've been just fine. Everyone's healthy, and we're all anxious for spring to really get here."

I knew exactly who called. It was my Aunt Nancy. My father had two brothers, Bud and Tom, and this was Uncle Bud's wife. Uncle Bud and Aunt Nancy had four boys: Ed, Dick, Bob, and Jim, and they were like brothers to me. She only called if there was bad news, or if they were headed our direction. That was always good news.

"I think that would very much be something he would like to do this summer. Would you like to speak to him?"

Diane called me to the phone, and said that Aunt Nancy wanted to speak to me. I quickly ran to the phone to see what she wanted. She invited me to go camping with their family to the Northwoods in northern Minnesota in June. She told me that to get to the Northwoods, we would travel from Chicago, Illinois to Duluth, Minnesota, then follow the north shore of Lake Superior to a town called Grand Marais. From Grand Marais, we would travel another sixty miles north on the Gunflint Trail to the end, and that was where we would be camping.

I was good with geography and had a general idea where Minnesota and Lake Superior were located. I tried to comprehend what she was saying and kept listening. She explained that their oldest son, Ed, could not go because of work and football. Her boys had taken a short three-day canoe trip last year, and they wanted to take another one this year, but they needed a fourth person. She asked if I would like to go.

I said I would love to go, and asked what I should bring. She told me to bring camping clothes, my fishing tackle, and pole. She explained that it was usually still chilly in June because the ice did not melt off the lakes until May. Warm clothes, including parkas and sweatshirts were important. Even though it was chilly, she said to bring a swimsuit in case we went swimming. I would also need personal toiletry items like a toothbrush, toothpaste, and deodorant. Since I was going in Ed's place, I could use his sea bag and sleeping bag, and share the tent with Dick. She mentioned someone named Rolf, the canoe outfitter at the end of the Gunflint Trail, who would rent us canoes and packs. I remember thinking that she was mispronouncing his name.

We finished our conversation, and I asked her how much money I would need. She said she would have her boys send a letter and review everything. We could plan the trip and the menu and get an estimate of the cost. She thought sixty dollars would be enough. After we were done, she put the boys on the phone.

Their voices were faint as they gathered around the phone in the kitchen. It had a five-foot cord that was usually twisted and hooked on the receiver to the spin-dial wall unit. Dick told me years later that the original cord was much longer, but it kept being pulled out because everyone stretched it as far as possible to get into another room. After the third time it was broken, Uncle Bud found the shortest cord he could find and used that one to solve the problem. Dick said he did the same thing later after he got married and his daughters stretched and broke their phone cords.

Our conversation was lively, and included the usual gibes about who was going to catch the biggest fish and the most. Dick kept saying how beautiful and wild it was, and how you could drink right out of the lakes. This was nothing like the state park camping I had done with them. Bob told me how Rolf showed them an unnamed lake near their campsite that he called Little Northern. Apparently, there was no real portage to get to it, but they pushed back through the woods and had a blast catching Northern pike. I asked who "Ralph" was, and Bob corrected me, saying the man's name was spelled R-O-L-F. I had never heard of that name.

Bob, usually the first to explain things, told me that Rolf and his wife, Gail, owned Way of the Wilderness canoe outfitting at the end of the Gunflint Trail. They rented canoes, paddles, tents, sleeping bags, packs, food, and anything else that might be needed for a canoe trip. You could be completely outfitted, or rent whatever you might need on a per-day basis. The boys described Rolf as "a real neat Norwegian guy with bushy eyebrows." They added that he and Gail were two of the nicest people. Nobody knew how long he had been up there, only that he seemed to know every nook and cranny of the wilderness.

We finished our phone call quickly because long-distance phone charges were a main budget concern in those days, especially for people with modest incomes. Dick was always good at following up details, and he promised to send a letter outlining the trip.

My mind was racing as I hung up the phone. Will I have enough time to get my money saved? Will I have to pay for a bus trip to Chicago, too? What will I tell my girlfriend? Where exactly is the Northwoods?

Diane's husband, Don, had the road atlas in his car, and he frequently had to work late at the YMCA. I decided to ride down to the gas station and see if they had a Minnesota road map. In those days, you could get free road maps at the local service stations, and they usually included surrounding states. I thought I remembered the Gulf station having a Midwestern United States map, so I jumped on my bicycle and went down to look. I had been to the Gulf station many times to put air in my bike tires or to buy a bottle of pop. They were closing as I got there, but I pushed the door open and went to the map section, where I found that I was right. There was a map that included Wisconsin, Illinois, and Minnesota. I grabbed it, and headed home. I could not wait to look at it and figure out where we were going.

When I got home, Diane watched as I spread out the map on the dining room table to see exactly where we were going. I located Duluth, Minnesota, and then followed the north shore of Lake Superior until I found Grand Marais. There was only one road heading north from Grand Marais. That must be the Gunflint Trail, I surmised. It ended in what looked like a massive matrix of inter-connected lakes.

"I guess this is where we're going," I said, pointing to the northeastern portion of Minnesota along the Canadian border.

Diane looked at the map with me and mentioned that Aunt Nancy and the boys seemed to have a great time camping there for the last two years. I supposed I would ride a Greyhound bus to get there, but I was surprised to hear

Diane say that I might be able to fly. The Head Office for the YMCA was in Chicago, and Don was able to get special passes to fly sometimes. She said she would talk to him when he got home.

I folded the map and ran upstairs to the third floor where my bedroom was, trying to remember how much money I had already saved from caddying. My snow shoveling money was gone, and I had just started the caddying season. I only had a few dollars, so I hoped the weather would cooperate, and I could make enough before June.

With every passing day, I tried to imagine what this canoe trip in the Northwoods would be like. It certainly sounded like the type of adventure every boy dreams about, and I hoped that the actual event would surpass the thrill of looking forward to it.

The time went by quickly with baseball season at the end of the school year and other things that happen the last three weeks of school. Weekends worked out perfectly, and I was able to caddy enough to raise money for the trip.

ON THE WAY

I was able to fly to Chicago when the time came, and I was very excited to fly alone as a fourteen-year-old boy. I heard that Chicago's O'Hare Airport was the busiest in the world, and I worried a little about my family being able to find me. I thought about seeing Aunt Nancy, Uncle Bud, Ed, Dick, Bob, and Jim. We had so much fun over the years, although I saw them less when they moved to Chicago. I actually saw them more frequently after Mom and Dad died. I would see them in the summer, and we spent time together because of the visits for funerals. This time was different, though. This was going to be a blast. I wondered if we were going to get the opportunity to stop at that new hamburger joint in Chicago. I really liked their French fries and milkshakes. I wondered how many hamburgers they had served.

When I got off the plane, Uncle Bud was waiting for me. His real name was Forrest, after his father and my grandfather. He came from work because it was closer to pick me up on his way home. Uncle Bud was about five feet ten inches, with blue eyes and thinning brown hair that showed gray streaks. He wore blue work trousers and a short-sleeved blue work shirt that matched. He always had the top button open, showing the upper portions of his hairy chest. He was very strong because he had worked in gas stations and did electrical wiring for Pullman cars at the railroad. He always had a big smile, and he greeted me with a hearty handshake.

"How are you doing, little Bill?" he said in a raspy voice.

I was named William M. Sanderson II after my father. Some family members called me Willy, but I preferred Bill, and Uncle Bud respected that. Uncle Bud was a kind but tough man. He loved to laugh and play cards, and he loved a cold bottle of beer. He had come to Chicago to learn the hydraulics business from a man he had known in Columbus, Ohio. The hydraulics business was okay for him. He made a living to support his family, and did on-site repairs for large customers. It was limited income, and required getting your hands dirty, doing design work, and troubleshooting. Raising four boys required money, but he did what he had to do to make it work.

Looking back, I think Uncle Bud worried about his family. The year before my dad died, their youngest brother, Tom, died of kidney disease. He left a wife and four children in an orphanage in Chicago. After my dad's death, Uncle Bud was the only brother left. With his own family's needs, there was little that he could do financially, but he always gave emotional support.

Uncle Bud broke my train of thoughts, and asked me if I was ready to do some fishing. I assured him I was.

"Where we're going, we'll definitely catch some fish," he said. "I just hope I leave enough fish in the lake for you and the boys to catch a couple."

"That's funny, I said. "Dick said the same thing. I guess I'll need to show you both up."

We had a good laugh and continued the visit all the way to their house. We talked about all kinds of things, but always seemed to come back to our camping trip. Like many Americans at that time, we went camping for vacation because that was all we could afford to do. The previous trips with my cousins had included camping, fishing, and swimming, but I kept trying to imagine what this new place with the great fishing would be like.

It seemed to take forever to get out of Chicago and out North Avenue to Wheaten where they lived. We finally pulled into the driveway, and I saw Dick and Bob packing the sixteen-foot aluminum fishing boat with the camping gear. I jumped out of the car and yelled hello. They returned the hello in unison.

Dick met me first. He wore blue jeans and a light blue denim shirt hanging loose out of his pants, and had on dirty and worn white tennis shoes. Dick was sixteen and a half at the time, and stood five feet and seven inches tall with curly brown hair. He was as strong as an ox because he played football and wrestled, and his arms were solid muscle with wrists that were like vice grips. He had a vertical forehead crease that deepened when he was concentrating on something, and it separated his deep blue eyes. Some people develop forehead creases because they frown in anger, and some get creases because of intense concentration. His was the latter.

Dick was not afraid of anything, although he occasionally complained about his back. He had a habit of jumping off garages when he was a kid. I also remembered the time he talked some of his friends into bending over a twenty-foot tree and launching him through the air to see if he could land in a tree twenty-five feet away. The launch was five-feet short of its destination. All we heard was the sound of tree limbs breaking before Dick hit the ground with a thud.

Bob greeted me next. He was fifteen years old, and came right behind Dick. He wore blue jeans, too, with a short-sleeved Banlon shirt that showed his arm muscles. His shirt was tucked, and his tennis shoes were not quite as dirty as Dick's. Bob was about five feet nine inches tall and was a gymnast so he had a triangular upper torso. He was very strong, and had the same deep blue eyes common to all the Sandersons. His brown hair had a blonde streak that went through the front portion of his bangs. I always asked if he colored it, but he claimed it was natural. Bob's wrinkles were around his mouth formed from smiling. He was a little quieter than his brothers were. Maybe it was because he was stung nearly one-hundred times when we were younger and rolled down the ravine and over a bee's nest at their house. He was the only one of us who rolled over the nest.

We all shook hands, greeted each other with smiles, and Dick grabbed my suitcase from the car.

"We need to get rid of this suitcase and put your gear in a sea bag like the rest of us," said Dick. "Let's go inside and get this changed out."

As soon as we got to the door, Aunt Nancy greeted me with a big smile and a hug, and made me feel at home by telling me to get cracking on switching my stuff to a sea bag, because she wanted to get on the road within the next hour.

Aunt Nancy was quite a character. She was about five feet seven inches tall, well endowed, and had auburn hair with freckles. She had gained weight since her days of diving as a teenager. Aunt Nancy came from a family of means. Her grandfather was Edward S. Matthias and he was elected to the Ohio Supreme Court and served for thirty-nine years. She was a rebel and did things her way. She loved to laugh, play cards, visit with family and friends, and was the source of many practical jokes over the years, and most of the time, she was the one who would start something. I always felt like one of the boys with her, and she made me feel like I fit in.

My youngest cousin, Jim, came from around the corner and said hello in a squeaky voice. Jim was thirteen, but had not yet developed physically like his older brothers. His voice had not changed yet. He was strong because he was a wrestler, and his blue eyes and wavy brown hair were his best features. As he got older, he had a remarkable likeness to Michael Landon. The shortest of the boys, he was only about five feet four inches tall, but grew a few more inches and filled out later in life. Jim loved to tease and loved to play pranks and he was just plain ornery. He asked about my trip, and I told him flying was fun, especially the take-off and landing.

When I got to the kitchen, Ed came around the corner and with a big smile, said hello, and gave me a big hug. Ed was the biggest of the cousins. He was five-feet eleven inches and weighed over two hundred pounds. He had a round face with blue eyes and short brown hair, neatly combed to the side. He played center on the football team, bass fiddle in a folk group, and did not look as muscular, but he was certainly strong in his own right.

Ed was always ready with a one-liner or a humorous story. He made me feel welcome, and today was no exception. He asked me about my flight, and then pointed out the hamburger and fixings he saved for me for dinner. As he told me to eat, he said I was going to love the Northwoods, and told this story.

ED'S STORY

"The first year we went, we camped at the campgrounds at the end of the Gunflint Trail. It was right along the portage between Seagull Lake and Saganaga Lake. It was the first time I ever witnessed water flowing north. Down here, the rivers flow south. We really had no idea what to expect; in fact, I don't even know how Mom and Dad found out about Gunflint.

We couldn't afford to rent a canoe, but Mom and Dad rented one anyway. The four of us weren't getting along with just one canoe, so Mom suggested I talk to Rolf and barter a canoe rental for cutting all the weeds in front of his building. That worked for Rolf, so I handled the scythe, and Dick swung the sickle for the better part of the day and we got an extra canoe. Jim was with me and was being a "lily dipper"[2]. The other two lily dippers, Dick and Bob, had the second canoe. Having two canoes enabled all four of us to get out on the lakes and fish.

The following year we took a three-day canoe trip. Rolf routed us through Seagull Lake and then north through Jasper Alpine, and Red Rock lakes, into Saganaga and back.[3] Rolf told us later that an hour after we had started out, some guys showed up on the ramp to Seagull Lake and claimed to have been lost for two days. Mom was now worried that she had just let her four boys ranging from ages twelve to seventeen go on a trip by themselves, and she asked Rolf if she'd lost her mind. Rolf assured her we'd be fine. He told me later that he could tell whether people were going to be able to handle a canoe trip or not. He told her we'd be safe. Besides, he could get to us in two hours time no matter

---

[2]A lily dipper is someone who does not place the entire paddle in the water when paddling and lets it drift along.
[3]See Appendix for map.

where we were by using a square stern canoe with a 3-Horsepower motor on the back. We made the trip just fine."

Sometimes it was hard to get Ed to slow down when he was telling a story, but I always enjoyed them, so I listened while he continued.

"A couple days after our canoe trip, I was in the store when Rolf came in and asked me if I'd be interested in going with him that evening out on Seagull Lake to check on a couple a campsites. He said he would be leaving about 7:00, and that I could go with him if it was okay with Mom and Dad. We wouldn't get home until around midnight, and I thought to myself that this was cool. I knew my brothers would be envious. I really didn't know what he meant exactly, but it really wasn't important. I was going out on Seagull with Rolf at night. Since I love to be on the water at night, I went.

The U.S. Forest Service had been purchasing all the private property in the Arrowhead area, and they were converting certain island sections back to campsites. If you look at a map of Minnesota, this northeastern section is shaped like an arrowhead, and that's the area the government sought to preserve and restore as wilderness.

Rolf said several prominent guys who loved the wilderness, William Magie, Sigurd Olson, Frank Robertson, and others, convinced the Federal government to return the wilderness to its natural state. The people who owned property in the wilderness area were bought out, and they had to move their cabins or have them burned. Rolf had a side job moving cabins for the Forest Service and burning lean-tos and other structures, and a ranger had asked him to check out a couple of sites that evening.

We jumped into his fourteen-foot aluminum Cadillac fishing boat with the 18-Horsepower motor on the back around 7:00 that evening, and scooted out to one of the small Islands on Seagull. Stepping ashore in front of a nice clearing,

Rolf checked it out for camping. Rolf told me all about the lakes and the people that used to live in the area and people who visited. He seemed to know everything about the area, because he used to have a cabin on Seagull Lake. He pointed to the area where his cabin was, somewhere on the north shore of Seagull Lake. He took it apart and moved it after it was sold. He added that the Forest Service was beginning to enforce the wilderness rules, which included horsepower restrictions on powerboats and restrictions on motors in certain areas.[4] He actually made me feel like an adult for the first time even though I was only seventeen.

On the trip back, Rolf had the throttle full open and I couldn't see a thing in the pitch dark. I was scared, but Rolf said he knew the bottom of the lake as well as he knew the top, so not to worry. He told me that he navigates the lakes by the skyline. Whether it's day or night, the skyline is the most reliable thing for dead reckoning. I love being on the water at night, but what Rolf was doing was incredible. I don't think he let off the throttle until we landed. You're going to like it up there, and you will really like Rolf. He is a neat guy. You will like his wife Gail, too," Ed concluded.

HEADING NORTH

Once Ed finished his story, it was a steady barrage of orders. Aunt Nancy was barking orders to everyone to get the boat and car packed for the trip. She told Uncle Bud to hurry and get cleaned up, and Uncle Bud was instructing Dick to put as much weight over the axle on the trailer and not so much on the tongue.

---

[4]The final regulations for the Boundary Waters were established on October 21, 1978, when the BWCA bill was signed by then-President Jimmy Carter. This bill finalized all the restrictions for motor- and canoe-only areas and formalized the name as the Boundary Waters Canoe Area Wilderness, or BWCAW.

Dick, Bob, Jim, and I took the gear from the house and garage and packed the boat and the trunk of the Uncle Bud's Ford. We put the clothes and dry goods in the trunk and most of the gear in the boat, and wrapped some items in plastic to keep them from getting wet in case it rained.

Dick asked Jim and me to give him a hand with the dining fly. Jim explained that they had used the thick canvas tarp for years in case of rain.

"Mom wants enough room under that tarp so the whole campground can have supper in case it rains. Wait 'til we have to put it up," he said.

Carefully we placed the thirty-by-thirty canvas into the boat. Uncle Bud finished his shower and came outside to survey the progress we made packing. He told us we put too much weight on the tongue and that we needed to shift the weight back further so it was over the trailer axle. With Uncle Bud watching, the four of us began moving and shifting the gear to change the weight distribution in the boat. It took a few minutes, but it looked better and he seemed satisfied.

Figure 1: Uncle Bud's Ford sedan

Uncle Bud had a full-sized Ford sedan with a trailer hitch that pulled the sixteen-foot aluminum fishing boat. The boat was powered by a 5.5-Horsepower Johnson motor hanging on the transom. The seating plan was that three of us sat in the back seat, and one of us would take turns riding in the front seat with Uncle Bud and Aunt Nancy. Every time we stopped, we would change the rider in the front seat. It was fun to be in the front seat. You could look at the road maps and follow along. I was glad that I brought my map from home so the backseat drivers could follow along, also.

Riding in the front seat was great because you could see the highways and signs better. I also had the opportunity to visit with Uncle Bud and Aunt Nancy. They always took an interest in my sister, Cindy and me. I think they wanted to make sure that we were doing okay considering our circumstances.

We drove long into the night until I heard Uncle Bud tell Aunt Nancy that he had to pull over and sleep for a while. Aunt Nancy instructed Dick to pull out the sleeping bags and told us that we could sleep outside on the grass.

"Where are we?" I asked.

"Somewhere in Northern Wisconsin," Aunt Nancy replied.

Bob took the flashlight and shined it over against what looked like a pasture fence, while Dick and Jim found the sleeping bags. He located a flat area near the fence, and we spread out our sleeping bags under the starlight. It took us about ten minutes to fall asleep, and we hoped it would not rain, knowing that Uncle Bud and Aunt Nancy would stay dry in the car if it did rain, but we would not be so lucky. We were so tired that even Uncle Bud's window-rattling snores did not keep us awake.

When the sun rose the next morning, we discovered that we had been sleeping next to a cow pasture. Cows were lazily grazing about thirty yards from us. The dew was heavy, and the air was crisp as we rubbed the sleep from our eyes. It smelled like a typical summer pasture in Ohio, a sort of grassy-green field scent. The sun was just beginning to stream through the distant trees and all of us seemed to awaken at the same time.

Dick returned from relieving himself and crawled back into his sleeping bag, grumbling that it was cold. He wiggled and grunted until he was settled in the warm bag. Just as we had finished our morning business and were rolling up our sleeping bags, we heard Aunt Nancy tell us it was time to hit the road. We finished putting the bags away, argued a little about whose turn it was to sit in the front, and piled into the car to continue our journey. As we came upon Duluth, I asked if we were heading up hill or if it was my imagination.

Uncle Bud explained that we were indeed, heading up hill, and that the elevations were higher than at home and would get even higher as we headed up the Gunflint Trail. He told me to notice that we were going straight up when we drove up the Gunflint Trail. It was so steep that you could see the entire town of Grand Marais below. Dick added that you could see the town silhouetted against the backdrop of Lake Superior.

The landscape was changing rapidly with Jack pines and birch trees becoming more and more noticeable. There were long-needled pines and aspens mixed in as well. The air remained crisp, but it was a beautiful day.

Duluth was exactly as I pictured it from my geography classes. Ore and grain ships, huge storage silos, and massive piles of iron ore dotted the shoreline. Duluth was the access point to the Eastern United States, and it was a small thriving city that had been there for years. History records that Duluth had more millionaires per capita than any other city in America in the early 1900s. Mining, shipping, railroads, and the lumber business gave Duluth its economic boost at the turn of the century. I could tell that next to shipping, the culture of the outdoors was the most important aspect of

Duluth. This became even more apparent as we traveled through the downtown area of Duluth. Everyone I saw wore flannel shirts and outdoors garb even though it was June. They seemed to be coming from or going to some outdoor activity. I wondered if I could be so hardy as to live in this region.

The ride from Duluth to Grand Marais was beautiful, following the North shore of Lake Superior. The road was elevated from the lake. There were steep drop-offs and cliffs in some places.

The shoreline was rocky, not sandy, and there were places to stop where you could climb down to the water's edge. The water was a lighter blue close to shore and got darker farther out. The two-foot waves splashed against the rocks and looked chilly. The terrain sloped up and away from the lake on the left as we traveled. Sometimes there were valleys and at other times steep cliffs with waterfalls and small creeks with water cascading down.

Every little town we passed through had interesting motels and cabins for rent. The souvenir shops were intriguing, and I wanted to stop at every one. We did stop at a couple and I bought some moccasins made of all leather with no hard soles. I could not wait to try them out.

When we reached Grand Marais, I found it fascinating. Aunt Nancy wanted to finish the grocery shopping, so the boys and I walked around town while she and Uncle Bud went to the IGA. She told us to meet at the entrance to the Gunflint Trail in forty-five minutes under the sign of the guy carrying the canoe. This longtime landmark of the Gunflint Trail is located along the main road from Duluth, and features a large wooden lumberjack with a canoe on his shoulders. It was about sixteen feet high and sixteen feet long. It was a great trail marker and is still there today.

We walked around the little town and went in and out of the shops along the shore of Lake Superior. We threw a few rocks in the lake because the water was too cold for swimming. I could feel the cooling effect that huge lake had on the shoreline. The cool lake water acted as a giant air conditioner, cooling the town with the breeze that blew from the South and West. I was glad I had on my parka and jeans.

The Gunflint Trail was actually a paved road in town, and Aunt Nancy and Uncle Bud met us by the entrance by the lumberjack just as planned.

Figure 2: Entrance to the Gunflint Trail

UP THE TRAIL

As we headed up the Gunflint Trail just outside of Grand Marais, we looked over our shoulders and saw the town below with the majesty of Lake Superior spread out behind it. The initial climb from town is quite steep enabling a view above the treetops. It was a sunny day, and the dark blue water provided a contrast behind the town with boats that dotted the harbor. I asked if we would see any bear or moose, and Uncle Bud replied that it was certainly possible.

"Rolf told us that moose are usually seen in the early morning and early evening, especially in marshy areas," Bob said. "Bear can be seen almost any time. Last year we saw deer and a fox on the trail, but no moose or bear."

The Gunflint Trail was paved in some places, like in town, but was generally a gravel road. Gravel roads led to lodges and businesses unseen from the trail and buried in the wilderness. There were signs marking their existence, though, so you knew which road to travel down to get to them. Rolf told us the Gunflint Trail originally started in the 1800s as a series of small roads and footpaths created to serve trading posts and prospectors. These early prospectors mined for gold and silver, and later they looked for iron ore. By 1900, the road ran from Grand Marais to Gunflint Lake to benefit some of these mining operations. In 1928, the road was extended to Seagull Lake with access to Saganaga Lake. That is considered Gunflint Trail's end.

Rolf later told us that Weyerhaeuser originally owned this area, but in the 1930s, a few adventuresome people tried to make a go of the tourist business and started lodges here and there. Electricity was installed in phases up the Trail in the 1950s. Rolf used a generator the first three years he was in business and received electricity in the latter part of 1959.

Gail spearheaded a book about the history of the business on the Gunflint Trail, called *A Taste of the Gunflint Trail*.[5] It contains the fascinating history of these businesses and wonderful recipes made famous by the lodges and destinations along the Trail.

As I looked at the map, it showed many lakes along the Trail, but few of them were visible as we traveled. I looked out both sides of the car, which was easy since it happened to be my turn to sit in the front on this stretch of the trip. The wilderness came right up to the road and was as thick as you could imagine. Jack pines, birch trees, long-needled pine, and aspens of various sizes and shapes seemed to cling to every

---

[5]Women of the Gunflint Trail, *A Taste of the Gunflint Trail: Stories & Recipes from the Lodges as Shared by the Women of the Gunflint Trail*. Adventure Publications, March 2005.

crevasse to survive in this rocky rugged environment that had little topsoil. Moss and lichens grabbed the rocks to hold their place. It looked as though this area had never been explored. In every marshy area we passed, I tried to see if a moose was grazing, but it was not going to be my lucky day. We passed several areas that looked like perfect places to see a moose, but we did not see one the whole trip.

At last we passed by the beautiful Gunflint Lake, visible from the road if you looked over your shoulder on the approach from the south. As we neared the end of the Trail, the road took several twists, and we saw a clearing on the right with several cars parked in the middle. We turned and drove through the clearing, and I saw a brown wooden sign with yellow routed letters that spelled Way of the Wilderness. There were bear paws imprinted on each of the upper corners, and I found out later that the bear paws were Rolf's trademark for the outfitting business.

We turned down a gravel road surrounded on the left side by a big boulder pile and a thick covering of Jack pines. On the right was a growth of birch and aspen trees whose leaves were shimmering in the cool June breeze.

CHAPTER 2 - MEETING ROLF AND GAIL

We parked the car and piled out when we reached our destination. We headed for the outfitting building, glad to be finally there. In the background, I could hear a dog barking from behind the outfitting building where it was chained. As we approached the red clapboard cedar-sided building, I could see a man in a billed cap with a flannel shirt and khaki pants through the large picture window. He got up from his desk and headed toward the door to greet us. That must be Rolf, I thought.

The dog I heard barking was an Alaskan Husky named Waysha. Rolf told me later that he never had a dogsled or dog team, but he used Waysha as the lead dog to tow the kids in their wagon.

Figure 3: Rolf standing by his sign

Waysha was the shortened form of a longer Indian name. Indian tribes like the Ojibwa and Chippewa had long names that told a story rather than using just one word to represent a thought or concept. Rolf learned several things about the language from Mrs. Powell, who was a full-blooded Indian. Waysha originally meant *a beautiful one*.

There were other names around the area that were native as well. Saganaga Lake actually meant *lake of many islands where it is easy to lose your way*. Sagnagons Lake meant *lake of many islands through the middle of the lake*, and Gabimichigami meant *lake with a river crossing it*. Rolf showed me a list of meanings that the lakes had in their Indian language. He said he began to understand more clearly how the language worked when he got to know the Indians he guided with when he first came to the Gunflint area.

Rolf greeted us at the door of the outfitting building with a warm smile and asked us how our trip had been. He was about five feet ten inches tall and had really bushy eyebrows like my cousins told me. I also noticed that he had slightly protruding upper eye sockets. He looked like he might be of Norwegian ancestry, and when he squinted his blue eyes were very visible when he spoke. He was not particularly burly or muscular, but he was rugged, strong, and seemed to be a perfect physical fit for the wilderness. His upper body was well developed from lifting canoes, and his upper arms were thick and strong.

He had warm facial features that made him look very pleasant, and I could tell he was a friendly sort, but not one to dominate a conversation. He usually waited to speak second out of politeness, although I would not say shy, just considerate and accommodating.

Aunt Nancy asked Rolf if he remembered the boys, and he said he did. She mentioned that I was Ed's replacement this year, since he could not come because of work and football. She was always sensitive to treat me like one of her own, and that meant a lot to me. As an adult, I realize how important it is to include people, and it seemed to come natural to Aunt Nancy. Little did I know that my

Figure 4: Outside new outfitting building

acquaintance with Rolf would be what it has turned out to be all these years later. He became my outfitter, my employer, and my friend, and our friendship still lasts to this day and over a span of forty years. She asked if Gail was around, and Rolf said she was over with the kids making supper and would probably stop in to say hello when she had a chance.

I turned my attention to the outfitting building. It looked to be about forty feet long and twenty feet wide, and was one and a half stories high. The upstairs loft doubled for additional storage and a sleeping area. There were steps leading to a small landing and to the front door, and the large picture window where I saw Rolf when we first arrived was in front and to the right.

In the middle of the main room was a large four-by-four wooden table, with built-in shelving under the tabletop. Maps were laid on top for trip planning, and beneath the tabletop, the shelves held maps for sale. The entire far wall was covered with connecting maps that showed all the wilderness area. On the right as you entered was Rolf's desk, and I saw that it had a few papers and the phone on it.

On the opposite wall was a print of a Canadian Mountie and an Indian running white water in a birch bark canoe, along with a map of the territory of Minnesota when the Quetico was part of the U.S.[6] I felt the inside of the building come alive with the wilderness. Rolf had built it from scratch to accommodate his needs.

Figure 5: Print on Rolf's wall

The four of us walked over to the map wall about the same time, and I immediately

---

[6]See Appendix for Minnesota Territory map.

tried to locate where we were. Dick read my mind and pointed to a spot on the map that Rolf had conveniently marked Way of the Wilderness. As I looked at this huge area of maps on the wall, its massive area dumbfounded me.

The wall maps included the wilderness region of Northeast Minnesota to the Canadian Border on the east and north also known as the Arrowhead section of Minnesota. You could buy a booklet of paper maps that covered the whole area, and they were made by the Fisher Map Company. Rolf was friends with Earl Fisher, and even used him for photography services. It was Earl Fisher and his photographer who flew in and took the picture for Rolf's original brochure.[7]

Rolf had taken the paper map booklet apart, and put the pages on the wall edge-to-edge to connect them in the proper order. The maps covered the area to the west just past Ely and forty miles north into what was known as the Quetico Provincial Park of Canada. At that time, the U.S. wilderness was known as the Superior National Forest and was under the administration of the U.S. Forest Service. It consisted of over one thousand lakes, and when it formally became the Boundary Waters Canoe Area in 1978, it included nearly one million acres including the lakes. The Quetico is almost one million acres also, and includes an additional 542 lakes, making the combined Quetico-Superior region over two million acres in size with more than 1,500 lakes.

The Quetico region alone extends sixty miles across and forty miles north and south, along the U.S.-Canadian border, and when combined with what is now the Boundary Waters, would make the "Quetico-Superior" three times the size of Rhode Island. The original inhabitants were the Indians, and later in the 1700s and 1800s the Europeans, particularly the

---

[7]These maps are still available today and have been updated throughout the years.

French, used the waterways in their fur-trading business. Known as the Voyageurs, they transported their furs by canoe from the very northern sections of the region down through the waterways and ultimately to Lake Superior.

Bob pointed to the map on the wall and showed me a three-day loop trip from Seagull Lake through Jasper, Alpine, Red Rock, and Saganaga lakes. They had taken this route previously, but I could not help but notice that it seemed like such a small area in comparison to the large map.

I also noticed that the foot trails, or *portages*, that connected the lakes were marked with an R. I knew from my geometry class that a rod was a measure of distance, but this was the first time I had seen it used. Rolf reminded us that a rod is sixteen-and-a-half feet, about the length of an Indian's canoe.[8]

Figure 6: Map of the Wilderness

I asked how they kept their maps dry, and Jim explained that the maps Rolf sold were waterproof. The ones on the wall were paper, but the ones on the shelving under the table in the middle of the room were the waterproof ones and were relatively inexpensive. I turned around to see Dick holding a can he picked up from the map table. It was an official-looking can, but I could not see what it said.

---

[8] Rods as units of measurement are outdated and typically only used in recreational canoeing for measuring portages on maps.

"It's dehydrated water," Dick said.

"You open the can and add a can of water."

"Yeah, right," I said, as I reached to read the label myself. I shook my head and smiled when I read the words on the can.

"Bernard's is the outfit that I get my dehydrated food from for canoe trips," Rolf said. "They gave it to me as a kind of a promotional item. I keep it here on the table. It's kind of funny to watch people pick it up and study it, because it looks so real with the printed label you know. Sometimes I ask people if they want me to pack some for the trail."

Rolf let out a hearty laugh, and Dick and I could not help but laugh, too, as I set the can back down on the table.

Rolf asked if we were planning a canoe trip this year, and we told him we wanted to try the Canadian side. He suggested a five-day trip up through Northern Light Lake, and used his finger to point out the route on the wall map.

Figure 7: Dehydrated water

Uncle Bud and Aunt Nancy both asked about the fishing this year, how bad the bugs were, and whether or not Clark Island was available for camping.

"Golly," said Rolf with an accent that sounded more like 'gully'. "They're catching Lakers [9] on Sag[10]. They're catching walleye and Smallmouth bass on Sag, too. Most of the fishing has been good all year."

---

[9]Lakers are also known as Lake trout.
[10]This was another name for Saganaga Lake, where we entered from the Gunflint Trail.

With a twinkle in his eye, he went on to say that he thought it might have something to do with the bugs.

"The bugs seemed to start early this year and have been with us right along."

"Mosquitoes?" asked Aunt Nancy.

"Oh, yeah, the mosquitoes, the black flies, and the no-see-ums," Rolf said. "I think they're feeding the fish. You'll catch fish."

Rolf asked if we were planning to stay out on Clark Island again, and Uncle Bud said we were. Uncle Bud liked that it was close and easy to fish almost anywhere. Rolf mentioned that there was an early-bird couple here at the end of May who was staying on Clark Island, but he was sure they came out the previous week. He told us there were several other campsites nearby if someone else was camping there already.

"I really like Clark Island and the blueberry bushes that are all over," said Aunt Nancy. "I think we added small blueberries to our pancakes every day last year."

Uncle Bud suggested that we take the canoes so we could paddle in tomorrow to plan our trip and pick up the other gear we would need. Rolf agreed and told us to meet down at the Saganaga landing where we could load. Just then, we heard a female voice and turned to see a woman with two kids coming toward us.

"Hi, Gail!" Aunt Nancy said with a big smile.

"Rolf said you would probably be here today," Gail said. "It's so nice to see you!"

Gail was five-feet and eight inches tall and had a slender build. She wore blue jeans, a sweatshirt, and tennis shoes. Most of the women in this area were bigger and generally heavier and more rugged-looking than Gail was. She had a sturdy look with strong hands and shoulders but she was by no means stocky. She had short brown hair, brown eyes, and a very warm and welcoming smile. She seemed like a genuinely caring person, the kind of person you would want for a nurse if you were sick. I found out later that she was a hard worker, loved to fish, and was not afraid of adventure.

I liked listening to Gail talk. She had a slight Minnesota accent because her Os were stretched out and tight, and she said "ya" instead of "you". Her accent was not as pronounced as many since she grew up in the West, but it was more noticeable than Rolf's.

Aunt Nancy asked if Gail remembered the boys and introduced us, calling us all by name and then introduced me, telling Gail that I was Ed's replacement.

Stanton and Stuart were Rolf and Gail's sons. Stanton was a special needs child who seemed to demand more attention from his mom and dad, but both were shy, and stayed close to their mom while she and Aunt Nancy chatted. Rolf and Gail had four children, and Aunt Nancy asked where Sandra and Susan were.

Gail said Sandra was back at the house listening for two-year-old Susan who was taking a nap. Sandra was quite a mother's little helper, and was waiting for her to awaken.

The Skrien house was actually a Quonset hut that Rolf purchased from the U.S. Forest Service in 1960 for five dollars on the condition that he move it. It was a twenty-foot by twenty-foot building that had three separate rooms and was located about one-hundred yards from the outfitting building. It was heated with oil, which made it easier to manage than wood. When he bought it, Rolf took it apart and hauled it over to this location using his pick-up truck on the ice in the winter. Large loads are easier to move in the winter when the lakes are frozen than trying to move things in canoes and boats. When the government re-classified the wilderness as a protected area, existing properties were purchased and the buildings were sold and moved or burned like Ed explained before we left their house in Chicago.

Gail hoped Susan was taking a long nap because tonight was movie night and the kids would be up late. The Forest Service usually showed a movie on the side of Rolf's outfitting building, and anyone who wanted to watch was welcome. We were told that the Skrien family, campers in the campground, and some of the kids who lived along the Gunflint Trail that went to school with the Skrien children would come. There was generally a nice crowd, and this was the first movie of the season, so the kids were excited.

Aunt Nancy suggested that we get moving if we wanted to get a bottle of pop and a candy bar before Cliff and Hilda closed the North Country Trading Post. She gave Bob some money and told him to pick up a block of ice for the cooler and meet down at the boat launch in fifteen minutes, then she continued visiting with Gail. We always thought Bob was Aunt Nancy's favorite, and that was why he always got picked to handle the errands.

We said our goodbyes to the grown-ups and took off running toward the cedar-sided building about one-hundred yards away. It sat at the top of a huge boulder just north of Rolf's outfitting building. It, too, was cedar-sided, and the inside was also made of cedar. Cliff and Hilda had a small, but complete selection of groceries. They also had fishing tackle, camping supplies, postcards, and souvenirs. Hilda greeted us as we tumbled into the store. Cliff, who was rearranging the merchandise on a shelf, turned around and said hello when he heard us enter the door.

Cliff and Hilda Waters had operated the store since 1956. They originally owned a place on Seagull Lake near Rolf's cabin. When Cal Rutstrum, a well-known author and Northwoods icon suggested that someone open a store and outfitting business at the end of the trail, Cliff and Hilda opened the store, and Rolf started the outfitting business. Cliff was average height, with thinning hair and a large forehead. Hilda was a short European woman from London and had a British accent. She was a little shorter than Cliff, and weighed around one-hundred thirty pounds. Cliff ran trap lines for years when he lived on Seagull Lake.

Their cabin was across the lake from Rolf's on Seagull Lake. They were both friendly people, and certainly knew the wilderness ways.

Dick told Hilda what we wanted, and she said we were in luck because the deliveryman came yesterday, and they were well stocked on pop. As I looked around the store, there was a rustic and outdoorsy feel to the whole place. The natural cedar walls with cedar shelving held a limited amount of merchandise, which made the store look larger than it was.

Figure 8: Icehouse built into the side of a hill

Cliff built this building, and I could see that it helped to be a carpenter if you live up here. Although it was not like the stores back home, there was just enough of what you might need in a pinch. The pictures of bear, deer, and moose made you feel like you were definitely in the wilderness, and it was a great place.

Bob said we needed ice and Cliff started out the backdoor leading to the icehouse. We followed him down a short path to a shed. When we got there, Cliff opened the unlocked door, reached into a pile of sawdust, and pulled out a twelve-inch square ice block. He brushed off the sawdust while holding it between the jaws of the metal ice tongs that were used for carrying blocks of ice.

Later, Rolf told us that in the winter they use saws to cut twelve-inch squares in the lake ice down to about six inches, and then tap them with a hammer. The ice blocks break all the way down to the unfrozen water, and then the blocks are floated to the shore. From there, the blocks are thrown or slid down a handmade ramp to sheds full of sawdust. Sawdust is such a good insulator that the ice stays frozen all summer long. A lot of the sheds are built into the side of rocks or drop-offs that help keep them shady and cool. Finding sawdust was never a problem with all the saw mills around. I was amazed at this Northwoods way of life, and thought cutting ice sounded like fun.

When we went back in the store, we all grabbed grape pop and candy bars, and paid Hilda. She thanked us as she rang up our sales on her adding machine, and the screen door slammed shut behind us as I followed the boys down the steep dirt road heading for the boat launch area.

Both sides of the road were dense with forest growth like the roads I saw all the way up the Trail. I did not know how an animal could get through there, and certainly did not know how a person could. I was familiar with state park camping where the underbrush was cleared and the only trees left were the big ones. Most state parks had gravel roads that circled around with numbered clearings for campsites. This was much different because every clearing was surrounded by thick untouched wilderness.

I took a deep breath and exhaled with a sigh, marveling about how fresh the air was and how good it smelled. Bob, who has hay fever, said his hay fever is never a problem here.

"Only the mosquitoes," he said, as he swatted a mosquito on his arm.

Dick mentioned that Rolf told them no poisonous snakes or poison ivy lived in the wilderness, but that we needed to be careful about ticks because they could be nasty.

When we got to the boat launch at the bottom of the hill, I was surprised once again at the rustic picture I saw. The boat launch was a section of dirt and gravel next to the road leading down to the water's edge. There were logs stacked on top of one another, and backfilled with dirt that acted like a dock to step out onto from your boat. You could launch on both sides of these twenty-five foot logs because the opening to the water's edge was a good one-hundred feet across. This section apparently had been cleared for boat launching because there were rock boulders on either side. I would have been surprised if it was this level and absent of rocks in its natural state.

TO CLARK ISLAND

About the time we got to the launch area, Rolf was pulling up in his station wagon with two aluminum Grumman canoes sitting upside down on the top. He had cleverly built wooden carriers for the top of the car and lined the tops with rubber inner tubes so the canoes would not slide around.

Uncle Bud also pulled up with boat in tow and prepared to back the trailer down into the water. He rolled the window down and shouted for us to unload the boat, and put part of the gear in the canoes. We could tow the canoes behind the boat.

While my cousins were scurrying around the boat and the car, I helped Rolf carry the canoes over to the water's edge. We set one of them in the water and set the other right next to it with the bows still on land so they would not drift away.

We loaded enough gear in the canoes so we could all sit in the boat, but Jim insisted that he was going to ride in the canoe. He claimed that there was not enough room for him in the boat with the gear and everyone else, so he stretched out in the last canoe, demonstrating how much room he would have.

We ran a long towline from the stern of the aluminum fishing boat and tied both canoes to it, one behind the other. Rolf asked Uncle Bud if he remembered how to get to our destination, and Uncle Bud said he did. He mentioned that the water was high with all the rain and melting snow, so we should not have any problem bottoming out in the narrows. He told Uncle Bud to have one of the boys sit in the bow to help navigate, and reminded us to stay clear of the silver rocks. The silver rocks were rocks that were near the surface and that had been scraped by aluminum canoes, leaving a shiny aluminum film on the rocks as evidence of the mistake.

We waved goodbye to Rolf and started our journey out through the bay, up through the channel to the narrows, and on to Saganaga Lake. I could not stop looking around at the thick wilderness that surrounded us right up to the water's edge. This was unlike anywhere I had ever been.

"This is God's country!" I said.

Uncle Bud had one hand on the arm of the outboard Johnson motor throttle, and in the other hand he had a lit cigarette that he was nursing as we traveled along.

"It sure is, Cousin Bill," he said. "I think you're going to like it up here."

As we made our way northward through the channel, I saw huge rocks and boulders spanning across the opening ahead of us. I heard Uncle Bud cut back on the throttle. The boat with the canoes and Jim in tow slowed down as we neared the narrows. Dick scrambled to the bow of the boat, and looked down into the water for submerged rocks near the surface.

As we started into the narrows, I thought there was no way we could get through it. The water was crystal clear, and you could see every rock on both sides and the bottom. The opening in some places could only have been five-feet wide and fifteen or so inches deep. As we snaked our way slowly through, we bumped a couple boulders, and Jim used a paddle to push away from his position in the canoe. The canoes in tow had a tendency to sway back and forth a little. Because the water flows north here, the journey out of the narrows is easier since you run with the current. It can be more challenging on the return run.

Somehow we made it through without hitting any rocks with the motor, mainly because Dick was pointing out the rocks on the right and left. As we pulled away from the narrows that lead to Saganaga Lake, Uncle Bud opened the boat to full throttle.

I remember Bob nudging me and pointing straight ahead, saying something like that was Clark Island and we should be there soon. It was hard to hear over the sound of the motor. As I looked in the direction that Bob was pointing, I saw an island with a solid stand of trees. I thought to myself that it did not look like anyone could camp there, but what did I know?

I was getting chilly. The speed of the boat created a face wind, and with the water temperature, which was still in the forties, it felt like a giant air conditioner was blowing at me. Although there was still plenty of light even in the early evening, the sun's warmth was not as effective this far north at this time of day. I huddled down a little to stay warm as we got closer to our destination.

Finally, I heard Uncle Bud cut back on the throttle as we made a wide left turn toward the right side of the island. He guided the boat with the canoes in tow toward an opening in the island where a giant boulder sloped down gently toward the water's edge. Unlike the shores of Ohio lakes that could be sandy or muddy, granite boulders form the most common shoreline here. Some of the giant rocks slope gradually, some are jagged, but most are a combination and require wading to land the boats or canoes without scraping and gouging.

Dick hopped over the bow as we neared shore with the bowline in hand, splashing as he stepped down in ankle deep water. He yelped as he felt the cold water, and pulled the boat to the shore. The sound of aluminum scratching the rock and the boat coasting to a stop meant we had finally arrived.

"Hey, don't forget about me!" said Jim, as the canoes coasted toward the boulder landing.

Bob grabbed the paddle as Jim extended it and helped guide the canoes in tow to a safe landing. I scrambled out of the boat, stepping into the cold water as well. Noticing how cold the water was, I grabbed the other canoe before it slammed into the rocky shore. As I pulled the canoe up, I slipped knee deep into the water. I made a loud comment about how cold the water was as I pulled the canoe to the shore.

"It's not time to go swimming, Bill," Bob said, laughing at my mishap.

When we got the boats secured, we scrambled around checking out the island area, trying to decide where we could pitch our tents. After a few minutes, we heard Aunt Nancy's voice as she barked out her usual rapid-fire commands.

"Boys, get over here and help us set up camp. Unload the boat and the canoes. Help us with our tent first, and then let's get the dining fly up. After that you can set up your tents. Bob, get the bug dope out of the kitchen box. These mosquitoes are terrible!"

The dining fly weighed a ton and took all six of us to set up. Uncle Bud tied his typical granny knots, unusual for a former sailor, and the rest of us tied hitches and knots to secure the dining fly for an extended stay. I was glad when that task was finished because it was the hardest job of the trip so far.

Uncle Bud started the first campfire and dinner was hot dogs and beans. How I love the fresh scent of a hearty campfire. Of course, dodging the smoke while the wind shifted was always challenging, but the smoke dissipates as the fire burns hotter. The fire was warm, and my pants were still wet from my boat landing episode. It seemed that our tennis shoes were wet the entire time we were camping. Getting in and out of boats and canoes all day gave little time for drying. Sometimes we got them dry at night around the campfire. Socks were easier to dry, but by the end of camping, most pairs were pretty rank.

By the time we set up our tents, rolled out the sleeping bags, threw in our sea bags full of our clothes, and climbed in, it was dark and cold. We were tired and worn out from our long day of driving and the excitement from our first day in the woods. The quiet of the evening with the absence of crickets, Cicadas, and other ground noise left only the occasional sound of a wolf howling, fish jumping, or the sound of a bird flying overhead. You have never really experienced complete silence until you have been in the northern wilderness.

The combination of all of these things made sleep come quickly. My evening prayers were short that night, but I thanked God for His creation and for family. We could hear the lonely call of the loons in the distance as we drifted off to sleep.

PREPARING FOR THE TRIP

The next thing I heard was Dick rustling to get out of his sleeping bag heading out to do his morning duty.

"Hey, guys," Bob yelled. "Right outside your tent are blueberries. I can reach them without getting out of the tent. Man, are they tasty!"

As Dick and I returned to our tent to get warm again from the chilly morning, we discovered the blueberry patch where we pitched our tents. The blueberries were small but tasty.

Aunt Nancy overheard our conversation about the berries and told us to pick a mess for breakfast. We gathered some and handed them to her to add to the pancake batter as she cooked the pancakes.

"Man, this is good eatin'," Uncle Bud growled, biting off another piece of bacon and shoving it in with a big bite of pancake smothered with Canadian maple syrup.

Aunt Nancy would buy Canadian maple syrup in silver one-gallon cans. It tasted great and lasted a long time even with six of us using it. Everything tastes better when you are camping. I did not drink coffee as a kid, but I always loved the smell and watching the grown-ups enjoy drinking it. Having a Coleman cook stove made cooking a lot easier, and the fire could be enjoyed for warmth. We all agreed that breakfast was good and devoured all that had been cooked.

Dick and I did the dishes that morning. I always preferred to get my duties done as soon as possible. For some reason, it seemed like getting work done first gave me more time to relax without worry. It might not make sense, but it always works for me.

Aunt Nancy told us that we needed to head back to the end of the trail and go over our trip plans with Rolf. She and Uncle Bud were going fishing and reminded us that the paddle in and back would take most of the day. She also said to take our money in case Rolf wanted a deposit, or if we wanted something from Cliff and Hilda's store.

I told Jim that I was going to use his dough, and he said, "Oh, really?" We had this dialogue where I would say, "Gimme your dough." He would always reply "Oh, really?" It originated from a Bill Cosby comedy routine and seemed to work for light moments. I also suggested we do some fishing on the way back to Rolf's, and maybe catch something for dinner.

After gathering our fishing gear and making some peanut butter and jelly sandwiches, we jumped in the canoes and headed back to the end of the trail to have Rolf map our trip. The winds were out of the north, and we were heading south, which made our paddle back a little easier. It seemed like no matter what direction you paddle, the wind is in your face. Once we got to the narrows, the wind was not as strong, so we made a few casts along the shoreline to see how the fishing was. Our hearts were not really in it because we were anxious to see Rolf and plan our trip. We paddled steadily, Dick and I were in one canoe, and Bob and Jim were in the other. Every once in while we would hear Bob say, "Quit lily dipping, Jim, and paddle."

As we neared the end of the trail boat launch, the race was on. Both canoes started a mad dash to see who would get there first and have the bragging rights for the day. We won by a boat length, as we both pushed nose-first onto the sand and gravel landing.

"Ha! Ha! We won!" Dick said, "Beat you to Rolf's!"

Dick took off in a dead run for the outfitting building with the three of us following in a dead heat. We arrived at the steps to the outfitting building at the same time. As we scrambled up to the porch, we heard Rolf's station wagon coming down the gravel road, and saw him wave to us out the window as he drew nearer. We exchanged our good mornings as Rolf walked toward us, and he asked if we had settled in okay. We told him we had, and Dick told him about setting our tents up in a blueberry patch.

"Oh yeah, the blueberries have been a little early this year so far. They usually are stronger in July and August. Gail has a secret spot she picks over by Seagull Lake. She gets nice messes and we have them on everything." Rolf said. "Come on in, guys. Let's talk about your trip this year."

Rolf pulled out a Fisher Canadian-Quetico waterproof map and began to point to the route with his weather-worn finger. He started describing a five-day trip that would take us into Northern Light Lake.

Rolf certainly looked like he had been up here all of his life. The khaki trousers, flannel shirts and faded green-billed cap he wore never covered his bushy eyebrows and blue eyes, or his Norwegian facial features.

He told us we would need to check in with Canadian Customs on the north section of Saganaga Lake, then head back south and then east to Northern Light Lake. He showed us where we could camp on the first night. Rolf used a black marking pen as he carefully marked the campsites with the letter "C." He marked the fishing spots with the letter "F," and told us where he thought the fish would be this time of year. In those days the maps did not have campsites marked on them.

The rules on the Canadian side have always been that you camp wherever you wish. Over the years, certain places are used more frequently and become popular campsites. Rolf knew where they were and always marked them for his customers. I might also add that even the ones that are now marked on maps were not always the original campsites that may have been used by the Indians and Voyageurs.

On the second day, he routed us to the Eastern bluffs of Northern Light, and suggested that we check out the Indian pictographs. He marked the map with a "P." Rolf told us that the Indians drew pictures frequently on the face of rock cliffs. They were usually pictures of hunters, moose, fox, turtles, and canoes. There were also pictures of Thunderbirds, and even a pelican. Rolf claimed that they still have not figured out the chemistry of the paint that was used, and how it has withstood the elements over the years. He told us that there were pictographs all throughout the Quetico and the Minnesota side, and he was not sure exactly the time when they would have been painted, or what was their purpose was. He thought they might be religious in nature.

He pointed out a walleye hole nearby, and encouraged us to catch fish for dinner. He marked more campsites for day two.

"You may not see anyone on the lake," he said, "so you should have your choice."

On the third day, he told us we would need to push up a creek that ran to the north out of Northern Light Lake. He told us there would be some beaver dam pullovers, and that there were no portages. He explained that we would have to get out and pull our canoes over the dams while standing on the clumps of grass and weeds beside the creek. If the beavers had been active, there may be quite a few more dams to contend with. He said the last time he was up there it was not too bad, but the mosquitoes would be thick, so we should dope up ahead of time.

Bob asked if Rolf still had some of the World War II Insect Repellant he called bug dope.

"You bet," he said. "You can buy a bottle for a quarter and it will last you for years. Remember you only need a little dab, then rub it all around."

Rolf shared how he bought several cases of Insect Repellant from an uncle in Minnesota who opened a military surplus store after the war. It was an oily yellow liquid in a small two-ounce bottle. The Army used it all over the world, even in the most difficult areas like jungles and such. A small dab rubbed in went a long way. I still have my bottle to this day. Rolf also bought rain ponchos for two dollars and rented them for ten cents a day. He had several items from this source that were great for camping.

Figure 9: World War II Insect Repellant

Rolf continued to route our trip, pointing out the long portage, which was about a mile long and located at the top of the creek. The portage ran east and west and would be the most northern part of the trip, but would enable us to start back toward home. He said it was a long portage, but not so much up and down. He suggested we stay on Mowe Lake on the fourth night, and continued to mark campsites and fishing spots. He also reminded us to stop at Canadian Customs on the way back.

"It's right on the way to Clark Island so that won't be bad," he said. "Besides, you can buy some of that Canadian candy while you are there. Does that sound like a good trip?"

We all agreed that it did.

Dick, the perennial fisherman, wanted to know what they were using for bait. Rolf told him that some of the guys used leaches, worms, or minnows for live bait. He usually took a few plugs and spoons that would catch almost anything. He said what you catch depends on where you are fishing. The Northern pike and bass would be in the coves, the walleyes in the rapids and holes, and lake trout would be in the deeper water. Rolf said fish always bite the same lures, no matter what kind they were. It just depended on where you fished for them. He reminded us to use leaders for the Northern pike since their sharp teeth could cut through regular line.

Rolf told a couple more stories about that area and pointed to the map as he spoke. He had guided up there for years and loved to tell the stories he remembered. Bob and Dick grabbed Rolf's marker and marked maps for themselves. I grabbed the map Rolf had marked and tucked it under my arm. Rolf asked what we might need for gear as he started toward the part of the building where the gear was kept.

"I know we will need at least three #4 Duluth packs and your camp half ax," Dick said. "That was really handy."

We followed Rolf to the gear area. The half ax was a thirty-two-inch handled ax that had a blade and a hammer arrangement like a hatchet, only larger, instead of having two blades like an ax. A half ax was lighter, compact for canoeing, and more versatile than the double blades.

Figure 10: #4 Duluth pack unchanged from the 1950s

The Duluth packs were invented in the late 1800s precisely for canoeing and portaging the canoe country. The envelope-style design allowed them to conform to the profile of the canoe, but not stick up too high to offer too much wind resistance. The canvas material was rugged, and the shoulder straps and tumpline[11] allowed you to portage heavy loads over rocky trails with certainty. I avoid using the term "ease" because there is not much that is easy about portaging in the Northwoods.

The portages are old Indian trails that were originally game trails linking lakes together. Since the terrain is rocky in some places and marshy in other areas, the portages are tough when they are dry, and particularly challenging when they are wet, rain-soaked, muddy, and littered with trees that may make obstructions when blown over by storms. A reasonable amount of strength and endurance are necessary; however, Duluth packs, padded yolks for canoe carrying, and equipment designed for the country make a big difference in manageability. Rolf bought his packs from a company called Duluth Tent and Awning Company, now known as Duluth Pack. They were made of canvas and were not only rugged, but easy to clean out.

Figure 11: Inside Rolf's building

Dehydrated food was just becoming available to the backpacker, and really seemed like the way to go, but the budget we had made it necessary for us to pack canned goods, like Dinty Moore beef stew and muscle our packs along. The rest of our gear was pretty basic as well. We had two brown canvas Army tents that had separate ground cloths. Our pots and pans were all metal camping items that

---

[11] A tumpline is a strap that goes around the forehead.

were left over from both of our family's supplies. Our clothes were in the green sea bags, and our sleeping bags were basic Sears and Roebuck camping items. We were young and enjoyed the challenge anyway.

The section where Rolf kept his equipment was obviously built for his needs and was kept very neat and tidy. He had a cabinet for cooking equipment that was originally in his cabin on Seagull Lake. He had shelves for the gear and had it organized by item. In the kitchen and food section he built small cubbyholes where items were neatly stacked. He told us that he was very fussy about packing for his people. He never wanted someone to get out on the trail and have an important item missing. This could spoil a trip and ruin a customer's relationship with him. Rolf also kept track of and filed information on each of his customers. He recorded who, when, and where each of his customers took their canoe trip, and what gear they rented. He later told me it helped him remember the names and faces of his customers, as well as the trips they took.

Rolf grabbed our packs, plastic liners for the packs, and the half ax, and then asked if we needed anything else.

"Other than this and the bug dope, that should do it," Dick said.

Bob pointed to the big green Old Town Canoe hanging from the rafters in the equipment section of the outfitters building. He then pointed to a Hamm's Beer light-up sign that was hanging on the wall. The sign featured a guy in a flannel shirt sitting in a red canoe similar to the one hanging from the rafters with a bent fishing pole in an area that looked like the Northwoods.

"That's Rolf in his red canoe, just like this green one," he said, pointing overhead to the green canoe.

"Really?" I said.

"Yes, that was me," said Rolf. "We took that shot over on Seagull Lake. I can show you on the map."

Rolf told us he did some calendar shots in the wilderness for Les Blacklock. Les used to do a wilderness calendar every year, and they had done several shots in a red canoe, not only for his calendars but also for the covers of the *Naturalist* magazine.

It was 1953, and Rolf was picking up his mail one day at Seagull Lodge when Jack Miles received a phone call from his nephew. Jack was originally from Wisconsin, but had lived on Miles Island on Seagull Lake along with other people who had places there. Jack's nephew, Art, worked for an advertising agency and wanted some outdoor photos. They were looking for someone to find some locations for their shots. Jack put Rolf on the phone and they set up a date to have a meeting with the art director at Seagull Lodge.

A few days later, Rolf met with the art director who had developed some sketches emphasizing the wilderness and was especially interested in capturing the deep blue sky. Rolf showed them where he thought they should do the shot, and arrangements were made for the actual shooting date. Rolf thought it was originally for a calendar, and they had told him that the shots they took might not even be selected but would be entered into a contest.

Before Rolf even heard about the final outcome of the contest, he was driving in Colorado the following year at Christmas time, and saw a whole billboard of the scene they had shot that day with him in the red canoe advertising Hamm's beer! Shortly thereafter, Hamm's used the photo for light-up signs to put in taverns, and they adopted the photos for their slogan, "Land of the Sky Blue Waters." The photos had been selected from the pictures entered in the contest. Rolf was sent a couple lighted signs, including one that had the water moving and rippling.

"Do you see how the pole is bending?" Rolf said, pointing to the light-up sign. "They had me tie a rock to my fishing line to make it look like I had a fish on. I told them I could probably catch one, but they said they didn't have time for that."

He laughed as he thought back to that day.

"They had all kinds of tricks," he said. "That was the hardest I ever worked on a guiding job. I don't think any of them had ever been in the woods before. They knew all the tricks of their trade, but they couldn't do anything for themselves up here, including going to the bathroom!"

Figure 12: Rolf's photo for Hamm's Beer

The one thing they did not do was touch up the color. They said the colors were so perfect and contrasting that they were just right the way they were.

The photographers were quite fussy about the color of the sky. Sometimes they would stop for several minutes to let a cloud pass so the sky would be without blemish, and the reflections on the deep blue water, just like the picture on the Way of the Wilderness brochure.

That picture was taken on Gull Lake near the outfitting building, and it is not touched up either. Rolf had a full-color photo of himself overlooking a lake while holding a paddle on his outfitting brochure there. The color is spectacular. He always said that this country is so beautiful, there is nothing you can do to improve it.

Rolf claimed that these ads for Hamm's Beer were the first outdoor beer shots that had been done. People were supposed to think "refreshing." He would often say, "The guy in the red canoe in the refreshing wilderness." After these ads began to run, other beer companies started to do outdoor ads as well. Rolf claims this national ad program helped the state of Minnesota. Minnesota became known to some as the Land of the Sky Blue Waters instead of the Land of 10,000 Lakes. People became more aware of Minnesota as a destination and the outdoor activities offered there.

Figure 13: Photo used by Hamm's Beer

It was much later in life that I asked Rolf about royalties. He told me that he received a percentage of each calendar sold by Les Blacklock. That became a nice income stream when he did the calendars.

He went on to say that he did not receive much at all from Hamm's Beer at first. However, when the Hamm's national ad campaign began to use Earl Hammond and his Kodiak bear named Sasha, Rolf was still guiding for the location shots. He later showed me checks he had saved from the royalties that he received when the ads ran on TV in the 1970s. He said sometimes those checks exceeded what he was making as an outfitter.

Rolf told us that he bought the Old Town red canoe from Seagull Lodge. He got the green one when he bought the cabin on Seagull Lake. He still has the original green canoe, but he painted it blue sometime later. It apparently came with the property. Both were beautiful Guide models with cedar planking and sold for about seventy-five dollars new. He said he did not rent them out because they were too heavy to portage. He seldom used them, especially since he purchased his Grumman lightweight canoe. It worked well for the photographs, though.

Rolf uses his lightweight Grumman canoe for trips now. It is durable, weighs fifty-two pounds, and is much easier to portage. Grumman made airplane parts during World War II. After the war, they modified their pontoon molds, opened the tops, and started making canoes. The canoes had a good design, particularly their rivets, and became very popular in the North Country. Some of their competitors who were used to making canoes for lakes without rocks, could not keep up with the competition. The rivets they used would sheer off on the rocks, causing leaks and all kinds of problems. Grumman pulled ahead of the other canoe-makers in the late forties and fifties.

Rolf enjoys telling the story about how he bought the canoe in 1946 for one-hundred and forty-six dollars. He still has the receipt. He told about heading to an outdoor store in Minneapolis to buy a pair of boots when he saw the canoe hanging in the store. It did not seem like the guy really wanted to sell it, but eventually he did. Rolf claims it was the most he ever spent for a shopping trip for boots.

He learned more about Grumman when he wrote to them about the canoe ten years later. He sent them the serial number along with some questions, and a few days later he received a letter telling him all about the canoe and how they modified the pontoon molds they used for the military airplanes and boats to make canoes. They had all the records and were quite informative. Rolf's was the second lightweight canoe shipped to Minnesota. When he brought it back from the store, the Indians made fun of it saying it would never last.

Rolf continued using Grumman for rentals, and ordered new canoes through Justine Kerfoot, who owned Gunflint Lodge and had the Grumman dealership. Rental canoes have to be replaced about every five years, so he would buy a few new ones every year to keep the inventory fresh.

He kept the lightweight canoe for personal use and rarely rented it. The only time he did was when every canoe he had was out and he really needed to satisfy a customer. He would borrow canoes from other outfitters if he ran out just to keep the business moving and to avoid renting the lightweight. With that, Rolf went over to his desk and added up the cost of the gear.

"You can pay me now for the maps and bug dope," he said. "We'll wait 'til you come off the trail to even up for the gear."

Rolf seemed to be sincere about wanting to hear about our trip when we returned. He had been all over that country for years, but always wanted to hear about something seen or heard that he might have missed, or just about what happened on your trip. It was no wonder people continued to come back to him for their outfitting needs year after year. He told me that seventy-five percent of his business was from repeat customers. We made a pop stop at Cliff and Hilda's, informed them enthusiastically about our trip, and were wished good luck by both of them as we headed out the door.

The paddle back up the narrows toward Clark Island was a little harder. The winds were still out of the north, and blowing steadily. We did not fish on the way back, since we were anxious to get back and start packing for our trip. Aunt Nancy and Uncle Bud knew we would be anxious to start our trip and figured we would want to prepare right away.

The night went fast, as we spread our groceries out and planned each meal. We carefully placed everything in plastic liners and packed them in the Duluth packs. Aunt Nancy reminded us to put the bread and eggs on the top so they would not get squashed. She also put in a slab of bacon and told us to eat the eggs and bacon first then eat the oatmeal. We had plenty of peanut butter and jelly for sandwiches, and enough grape Kool-aid to turn the lakes purple. The Dinty Moore beef stew would work for our dinners when there was no fresh fried fish.

You could still take cans into the wilderness then, but cans were banned a few years later. You cannot take glass anymore either. Everything you take, you must pack out or burn if you can. Rolf taught us to leave no trace of having been there. This "Leave No Trace" rule has now been formerly adopted as the standard for travel in the BWCAW.

There were also no toilets in the wilderness at that time, and it is still that way today on the Canadian side. The American campsites have a fire grate and a toilet box over a dug hole. The rangers check these privies regularly and relocate them when they are full. The Canadians allow camping anywhere, but the American side requires camping only on designated sites marked on the maps.

After the sun went down that night, we star-gazed for about an hour, looking for shooting stars in the most incredible starlit sky I had ever seen. We finally got tired of swatting mosquitoes and headed to the tents. Tomorrow we would start, and I could not wait.

As I started my evening prayers, I took a moment to thank God again for the chance to be here in this place and asked Him to keep us safe on our journey.

CHAPTER 3 - THE FIRST TRIP

The next morning met us with an overcast sky that made the water look gray. The winds were westerly but not too strong. Since we were heading north to check in at Canadian Customs first, the wind would be at our side and not in our face.

Aunt Nancy made another pancake breakfast with bacon, and we gobbled it down. Bob and Jim did the dishes and we tore down the tents and packed our gear in preparation to leave. We finally shoved off around 9:00 a.m. with Aunt Nancy and Uncle Bud watching us and reminding us to be safe and use common sense.

Figure 14: Seaplane outside Canadian Customs

The paddle to Customs was not too bad, and only took about an hour and a half. It felt good to get out of the canoes and stretch when we pulled alongside the docks and tied our canoes.

The camping pass was only a dollar a night, so four dollars covered it. The ranger gave us a sticker for the canoe that served as a license after we showed him our trip plan so he knew what to charge us and for safety concerns. The rangers kept track of people in case there was an emergency like a fire. Knowing who was in which area was a good safety precaution. Dick was the only one who needed a fishing license because he was over sixteen, and he paid for that. After we bought some pop and candy bars and finished eating them, we headed back to the canoes to continue our trip.

The paddle south was about the same as the paddle north, and the trip was uneventful. The winds were now westerly, and not in our faces. The clouds began to break a little by lunchtime, and the sun peaked out. Seeing the sunshine on the lake was beautiful, and it made us feel good on the inside, too.

Clipping along on a sun-brightened lake with the shimmering reflections off the water was exhilarating, and very different than what if feels like to move under gray skies. It was nice to see the sun and we stopped for our lunch on a little island along the way.

As we worked our way into the first portage, Bob pointed and said he thought it was straight ahead of us. Rolf told us this portage used to have a railroad track on it. The guides who had fishing parties used to meet here and help each other use the tracks to portage the boats with motors over to Northern Light Lake. It was much easier to portage with a couple guides than by yourself. They would meet again on the other side to help going back.

Dick asked Bob if this was where Rolf told his coffee story, and Bob said he thought it was on one of those islands we already passed.

"I missed that story," I said. "What was it?"

As we paddled along, Bob told me the story about Rolf and his coffee.

Rolf had taken two couples out for a day of fishing. They were from Chicago and seemed a little uncomfortable about being out, but were doing okay. He met one of the other guides for lunch on an island somewhere near this portage.

Rolf usually started a fire for lunch so he could make fresh coffee. He had a #10 can with a bailing wire attached for a handle, and he would dip it into the fresh lake water to get water for the coffee. He would heat it up, drop in the grounds, and cook it for a bit, then he would add a little cold water to settle the grounds to the bottom, and pour out the fresh brew. He said everyone around raved about his coffee, even though they could not believe they were drinking coffee out of a can.

Apparently, one of the women in the group wanted nothing to do with the lake water, so the other guide who was with them said he would get water from a spring on the other side of the island. He jumped in one of the boats, went to the other side of the island, and when he was out sight he filled his bucket with lake water and brought it back for Rolf to use for the coffee. The woman raved about how good the spring water was. The day was saved and Rolf's coffee was still the best.

"Rolf must have a ton of stories," I said.

It felt good to get out of the canoes and portage because you can get stiff from being crammed into the canoe. Getting out to stretch feels good. The portage was not bad and we pushed off to look for a campsite, and when we found it, we set up camp and headed out to fish. We caught a few, but nothing big enough to fry, so the beef stew we brought was tasty and sleep came easily after a long day on the water.

The next morning we had a breakfast of bacon and eggs, and then headed for the pictographs in Trafalgar Bay. It took us a while to get there and they were hard to find, but we finally did. Rolf said we probably would not want to take pictures of the paintings because some of the Indians think there are still spirits in them, so we just looked at them and moved on. We were anxious to do some fishing in the afternoon and started looking for a campsite. We found a good island, set up the tents, and headed out to fish. This time we caught a nice mess of fish and fried them up for dinner.

As the evening wore on and Jim finished chopping all the firewood we would need for the night, he then began chopping something in the logs we were sitting on. Using large ten-inch letters, it looked like the start of "Sanderson Island." Each of us took turns with the half ax, and when we finished, we had created an island marker from the log. We took pictures of it and included the moose head skull we found on the island.

We had a great laugh over it, and Sanderson Island was a hit with us.

Figure 15: Dick & Bill on "Sanderson Island"

The next leg of our trip was to head north up the creek, make a turn west, and head south at the northern tip of our loop. We remembered Rolf telling us that there would be some beaver dams along the way, and we would have to pull over them, so we were anxious to get going.

Dick reminded us to fill our canteens in the open water before we got into the creek. Rolf always told us to avoid the water in the creeks and around the beaver activity, and drink the water in the open lakes and away from shore instead. This would be the last open water until we made it to the top of the creek and over the portage. Rolf cautioned us about drinking the wrong water because we could end up with the beaver fever[12]. We also remembered to put on the insect repellant before we got into the creek.

---

[12]Beaver fever is an intestinal infection caused by eating or drinking contaminated food or water.

The beavers had been active as Rolf said they might have been, and no sooner had we put into the creek, we had to get out of our canoes to push and pull over the log jam of sticks, branches, and small trees in our way. There was no easy way to do it except to trudge through the squishy grass creek bank pulling, lifting, and sliding our way along. In some places we were knee deep in water and mud, and the bugs were definitely active. In between the dams we were able to paddle short distances, only to be faced with pulling and tugging over another. In several places, it took all four of us to pull one canoe.

Unfortunately, the goal we were striving for was a mile-long portage at the end, and when we finally reached the portage, we were not sure it was the portage because it was so overgrown.

Figure 16: Canoeing through a beaver dam

"This is it," Bob said. "See the moose hoof tracks in the mud? All these portages are still game trails according to Rolf. The moose are the only ones who have been up here so far this year!"

We discussed whether or not we should walk it first to see if it was passable, but we decided not to, and broke out the food for lunch. We were all pretty tired after the morning workout and lunch under the sunny sky was refreshing.

"I'm gonna tell Rolf that those weren't beaver dam pull overs. Those were *damn* beaver pull overs," Dick said with a laugh.

"You're right, Dick. We'll have to tell Rolf," Bob agreed.

After a lunch of peanut butter and jelly sandwiches and cookies, I asked again if we should walk it first as I hoisted a pack on my back, but nobody seemed interested.

"I'm not walking it first," Dick said, hoisting the canoe on his shoulders following me through the woods.

That portage was every bit a mile long, and there were several downed trees, but it was passable. The bugs were still thick and there were a couple of soft areas, but the moose and other game had done a good job of tromping down the trail.

I saw the water through the trees, and we knew we had finally reached the end. The first sight of blue water peeking through the tree limbs was always a welcome sign and signaled the end of the portage.

Dick bragged about making the whole portage without putting the canoe down, and we all gulped down water as soon as we dropped our gear. Bob and Jim were right behind us, but that portage seemed to take forever. It was another mile walking back and another mile with the return trip and gear. As we finished reloading and settled back down into the canoes, we decided it was time to start looking for a campsite.

"Let's find something quick and get away from these bugs and out of this wind." Bob said, swatting and smacking as he spoke.

Rolf had marked a couple campsites and we started our paddle in the direction of where they were. We found one on an island, and pulled up for a look. It was not as nice as Sanderson Island, but it would do. We were all pretty tired, and no one moved very quickly, but we unloaded the canoes, worked on getting dinner, and scouted out where we should pitch our tents.

After supper, Dick and Bob decided they wanted to fish, but Jim and I decided to stay at camp and keep the fire stoked. We thought we might cast a little from shore, but we were too tired to go out in the canoe.

All of a sudden, we heard a commotion and Dick yelling from inside his tent.

"What's wrong?" I shouted.

"That doggone bottle of apricot brandy Mom gave us spilled all over my clothes inside my bag!" Dick said.

The rest of us laughed at Dick's dilemma, which kind of annoyed him.

"Here, give me those clothes, you lush," I said, laughing. "I'll string a line and hang them up to dry while you and Bob fish. I'm not going to wash them, but they won't smell too bad when they dry."

"Is there any left, or did you drink it all and spill it all?" Jim said, giggling.

"There's half of a bottle left," Dick said, as he pulled the bottle out of his back pocket and held it up for us to see.

Aunt Nancy always sent a small bottle of brandy with us for our trips. She thought it would warm us up on cool evenings. She also thought it would be medicinal should someone get sick or hurt. We always forgot about it since none of us had ever really had any alcohol as kids.

"Actually that might taste good tonight after a long day, especially since it seems to be getting chillier," Bob said, digging out a clothesline from the Duluth pack and handing it to me.

We strung the clothesline between two pine trees near the fire, and Jim and I did the dishes while Dick and Bob took off to fish.

About an hour later we heard the canoe scratch the shore and heard the guys talking as they pulled in and stowed the canoe.

I asked if they had any luck, and Dick said they had not.

"It was really getting windy and the canoe was hard to control," Bob said.

We wondered if some bad weather might be coming in. Usually in the Northern Hemisphere, the wind settles down in the evening unless a front is moving through. The four of us sat around the fire that night and each of us had a sip of the apricot brandy. We were all tired when we turned in as the clouds and rain came.

Rain is easier to deal with at night when you are dry and inside your tent. It is more tiresome when it rains while breaking camp. Luckily, when we awoke, it was only drizzling, but the wind had continued to build. We got on the water after breakfast and paddled down to the first portage. It was not an easy paddle with the wind, and took us most of the morning.

"We're going to have to lash the canoes together if we plan to travel today," Dick said as he looked over the gray lake.

He instructed us to find at least two trees five inches in diameter and about ten feet long. We would lash the canoes about two feet apart and it would make them more stable in the water, similar to a pontoon boat.

Dick had been in the Sea Scouts for a couple of years and learned some of the tricks and tips we used from there. Rolf later scolded us for lashing the canoes together, and said there was a danger of capsizing when you do.

We scouted around and found four small trees five inches in diameter. Dick trimmed them to the right length and laid them in position. He carefully placed the small trees across each canoe and lashed them together using clove hitches to tie them to the cross members of the canoes. He grabbed each lash point and tugged to make sure it was fastened tightly, and after he finished with the last one, he stepped back to survey the finished product.

"I think that should do it. Let's get loaded and go," Dick said.

As soon as we shoved off the portage, the rain stopped, but the wind persisted. Luckily, we did not have any portages, so we did not have to take things apart. The waves were small, but paddling into the teeth of that strong westerly wind was a lot of work. We stopped for lunch and fished a little, but fishing was somewhat tricky with the canoes being lashed together.

When we finally found our final camping place, the rain stopped. We were all wet and uncomfortable. Luckily, the lashed canoes worked, and instead of being wind bound for a day, we made a travel day of it. We unloaded the canoes and set up camp, then fixed the final dinner meal in our pack.

I walked down to the landing area where we beached our canoes to rinse out my cup, but when I got to the water's edge, I noticed that the canoes were fifty yards off shore and drifting away. We forgot to pull them up on shore, and when the weight was removed from them, they became buoyant. The wind and wave action worked them loose, and they started drifting away.

Losing canoes was a somewhat common occurrence. Storms can blow them into the water or people forget to pull them ashore, like in our case. Retrieving them can be tricky, though, and sometimes you have to hike around the shoreline through the woods to retrieve them. Since we were camped on an island, I knew we might need to find some trees and build a makeshift raft. After this mishap, we always made sure to pull them completely out of the water and turned them upside down.

"Guys! Get over here quick!" I shouted. "The canoes got loose!"

Everyone scrambled to the water's edge and started laughing, each blaming the other for not pulling the canoes up on shore. Then we looked at each other, wondering who was going in after them. Before we could decide, Jim had his shoes off and was peeling off his sweatshirt, ready to jump in. Dick handed him a life jacket and after a little arguing about whether or not to wear it, Jim put it on. It did not take long to reach the canoes, although watching him swim in that cold water had us all worried. We felt more relieved with each stroke.

When he got to the first canoe, he slid over the side of it and was able to roll in without tipping it over. Since they were lashed together, they were more stable and easier to get into. Getting in a canoe from the water takes skill and agility. Rolf required first time canoeists to practice getting in a canoe that had been swamped. He had them swamp a canoe near his dock and practice getting into it. He even made us do it in June with the water temperature in the fifties.

Behind us someone was yelling.

"Hey, Jim!" Dick said, running toward the water, camera in hand. "Get back in the water so I can get a picture!"

Jim refused to get back in the water, claiming that it was too cold. He also discovered that there were no paddles in the canoes, and was trying to figure out how he was going to get them to shore.

We looked over to the spot where we landed the canoes, and there were all four paddles on the ground by a big pine tree. Bob was yelling something about swimming them back, but it was hard to hear over the wind and the small waves lapping against the side of the canoes. The next thing we saw was Jim laying on the bow, dog-paddling like mad to get back. It seemed like it took forever for him to go fifty yards, but he finally made it.

We helped him get into dry clothes and sat him by the campfire to keep him warm. The wind continued to blow, making it feel chilly.

"A little of that brandy might taste good," Jim said, laughing.

"We still have some. I'll go get it," Dick said.

"If we're out, we can always suck some out of your clothes," I added, while Dick was heading to the tent to get the brandy.

"I guess this is what Mom was thinking when she gave it to us," Bob said.

We all had a couple of sips of brandy while we replayed the mishap and laughed again at our near misfortune. We especially enjoyed Dick telling Jim to get back in the water for a picture, and agreed that this would be a story to remember.

The wind finally began to settle down that evening, although it helped us dry our clothes a bit. As we sat around the fire talking about our trip and all that we had done, we were visited by a couple chipmunks that scurried around us looking for crumbs. We had fun watching them dart and scamper as we threw them food scraps. It was chilly when the sun went down, and the stars began to shine through the breaking clouds showing the vast beauty of the night sky.

The fire was warm and our spirits were high. There is always a conflicting feeling about the last night out on the trail. Part of me wanted to stay, but I looked forward to a bottle of soda and a candy bar and the opportunity to return to civilization and interact with others.

We went to bed and woke up the next morning to a sunny sky with westerly winds that were not too strong. We made our last portage and the long paddle to the Canadian Customs. We checked in with the same ranger we saw five days earlier, and talked a little about our trip. The sodas were great, and the candy bars hit the spot.

We jumped back in the canoes and headed to Clark Island where Aunt Nancy and Uncle Bud were still camped. Aunt Nancy saw us coming across the water and started shouting and waiving.

Uncle Bud wandered down from the campfire just as we pulled in, and said it was great to see us.

"Did you catch any fish?" he said.

"Oh yeah, we caught fish!" we said in unison.

"How have you two been?" I asked.

"We've been fine," said Aunt Nancy. "Fishing every day and napping in the afternoon. Rolf came out one night. He said he was going to try to get here tonight if he could, and hear all about your trip."

CHAPTER 4 - ROLF COMES TO VISIT

We unloaded our canoes and spent the afternoon setting up our tents, drying out our clothes, and telling about the trip. Aunt Nancy especially liked the story about the canoes getting loose. Uncle Bud caught some walleye and bass that morning, and Aunt Nancy fried potatoes to go with the fish. She also made cornbread and corn.

The meal was fabulous. It was better than the stew we had been eating for five days. Just as we finished eating, we heard Rolf's motor in the distance. He always had it full throttle and made good time heading in the last quarter mile to our campsite.

"You made it back, I see," Rolf said, stepping out of his boat and heading up the hill to our campfire. "I can't wait to hear all about your trip,"

Uncle Bud offered Rolf a beer and Rolf accepted. Uncle Bud reached into the cooler, pulled out a Hamm's Beer, put two triangular holes in the top with the can opener he called his church key, and handed it to Rolf.

"Hamm's Beer, that's a good choice." Rolf laughed as he took a sip. "You know if you run out of ice, you can just put these in the water. They'll stay plenty cold."

"I know. I've got a six pack in the lake now," Uncle Bud said with a big laugh.

Rolf said he used to keep canned food in the lake under the ice so it would not freeze when he lived up here in the winter in the cabin over on Seagull. Sometimes the labels would come off, and he was not always sure what he was opening. For instance, he would think he was opening green beans and it would be sweet potatoes. He said beer is a little easier to identify with the printing right on the can.

It was a beautiful evening and just perfect for sitting around the campfire and telling stories. The bugs were a little bit of a nuisance, though. We told all about our trip from start to finish. We told about the fishing, claiming Sanderson Island, the beaver dam pullovers that we renamed the damn beaver pullovers, the spilled brandy, and we ended with the story about the canoes drifting away. That was when Rolf asked about how we lashed the canoes.

He calmly explained like the caring instructor he was but with a gentle scolding, that it is not a good idea to lash canoes together. In choppy water, a wave can drive one side of the canoe down into the water, which would cause the other side to stick up. This, in turn, could cause you to swamp. A single canoe is buoyant and can pivot at the center and roll with the waves. They are not as able to pivot when you lash them together. Lashing is okay in high winds and on flat water, but not in wave waters. He did not recommend lashing canoes together.

We talked on about our trip, until I asked him what brought him to this area. The two-hour response to that question was captivating and riveting.

ROLF TELLS HIS STORY

"I was one of seven and was born in Ulen, which is in northwestern part of Minnesota. I played some sports and had an interest in drama and music. I also liked science and biology. When I was in high school, I worked part-time as a bellhop, at a dry cleaning shop, as a farmhand, and even did some driving for a veterinarian. When I graduated from Morris High School, I went to the West Central Agricultural School in Morris for three years. The school was close to my house. I continued my studies in the sciences, added physics to my list of courses, and generally enjoyed studying business and business law.

After that, and before World War II really got going, I spent a little time in San Diego where I attended a vocational school and learned to do aircraft inspection. Next, I joined the U.S. Coast Guard, and spent three-and-a-half years there. We mainly patrolled the North Pacific, and toward the end of my time, we patrolled the Kurile Islands.

After World War II, my brothers and I tried to buy a bank in Morris, Minnesota, but we were outbid. Those were the days when buying a small bank was possible. My father, who was a bank examiner, gave us the idea of buying the bank, and even though we missed out, he later became a vice president of that same bank.

The first time I ever came up to this part of Minnesota was in 1928, when I was seven. We were taking a family vacation and were going to Thunder Bay, which was called Main Bay in those days. We were sightseeing and camping along the way. It took awhile to travel in those days, because the roads were still gravel and full of ruts. They were just beginning to pave Route 61 from Duluth to Two Harbors, but had not finished. It was still gravel from Two Harbors north to Thunder Bay.

We camped at Tofte and then went on to Fort William and Port Arthur. We were going to Mineral Center, and we had to cross over a wooden bridge they called the Outlaw Bridge to get there. I'm not sure exactly why they called it the Outlaw Bridge, but I remember what it looked like. I remember that bridge because it was made out of timbers and built in three sections—up, over, and then down.

I'm not sure if it was more than one tribe, but a fair number of Indians lived there in Mineral City. I can still remember the teepees and layout. They'd sell Indian jewelry and items they had made, and I especially liked the hard maple sugar candy. As a seven-year-old, that was my favorite thing. We'd been camping along the shore of Lake Superior and eventually ended up in Grand Marais. Grand Marais was small back then, smaller than it is now. The streets were cobblestone and gravel, and the sidewalks were made of rough boards.

We stopped at a little Standard Oil gas station with an old-fashioned canopy over the gas pumps. It had just rained, and the depression that formed in the gravel by the pumps had collected water, right next to where you step out of the car. Of course, in those days, you traveled all dressed up even when you were camping, and when I jumped out of the car, I stepped right in the middle of the puddle with my new shoes. Dad gave me heck, and that was my first experience in Grand Marais.

We asked about going up the Gunflint Trail, and we did go a couple of miles up, but it seemed like everyone we asked discouraged us unless we had extra gas, extra tires, extra oil, and plenty of food. It was almost as though people were afraid of the area. We never made it up here at that time. We went up to the Hedstrom's Mill area that was clear, and they were growing potatoes up there. They'd cleared the rocks away to make it a field. We went as far as the pines, saw them, and then came back into Grand Marais. We didn't make it very far, you know.

It wasn't until the spring of 1946, that I actually made it up here to camp and fish. After the war, three of my buddies and I planned a canoe trip up here and came up in April of 1946. We'd all just been discharged and we wanted to do something fun. We stayed in Two Harbors the first night and made the rest of the trip the next day. When we got here, there were still ice chunks on Seagull Lake. The country didn't have a name then. It wasn't called the Boundary Waters or anything special. It was just wilderness or the North Country.

We brought our own equipment, old military gear, and one canoe. Our canoe was lightweight. A friend of mine who worked for Piper Cub got us a bolt of airplane fabric, and we recovered and soaked the whole thing. It was almost like an eggshell and only weighed about forty-two pounds. We rented another canoe, a heavy eighteen-foot canvas Old Town, from Russell Blankenburg, who owned End of the Trail Lodge and Seagull Lodge.[13] Russell was getting ready for the May 1 trout opener and his equipment was available.

As we started out on Saganaga Lake, we heard a noise like someone was pounding nails. We stopped to have a look and met this guy from Des Moines, Iowa, who used to be a detective there. He wanted to get away from it all and was building on this Island. We got to talking to him, and he said he'd been on a trip up into the Canadian side called the Quetico and had been up to Kawnipi Lake. That sounded real fine to us so we said we wanted to go up there. We had a small map that showed the general area, but didn't have any detail. It was only the size of a sheet of paper. You couldn't canoe from it because it was too general and small scale, so he handed us a map and told us we better take it or we would be lost for sure. It was a small map, too, but really detailed. It showed the area well. Without that map, we couldn't have gotten past Saganaga.

We took off and because it was cloudy, we thought we were heading west but we ended up heading north into Cache Bay. There was a ranger station there to check people into the Quetico, but it was early in the season and it was still locked. The cabin isn't where it is now. It was at the opening of Cache Bay so it would be easier to see since most people came up from Ely before the road was built up here.

We got mixed up a little trying to find Silver Falls, which is easy even with today's maps. We went down the wrong opening. We could hear the falls but we were in the wrong bay. Anyway, we finally found the portage, and one of the

---

[13]End of the Trail Lodge was called Saganaga Fishing Camp at that time.

guys dropped a line in where the water pools into the river there. Wouldn't you know he caught a two-and-a-half-foot Lake trout on the first try? That was way the fishing was back then. We caught fish everywhere we stopped that trip, especially up in the Falls Chain. They were thick in there, because most fish spawn in the spring. Lake trout are the only ones that spawn in the fall. You could just throw your line out, and if you missed it, was your fault. I remember one of the guys caught six in a row on six casts. We filed the barbs off our hooks and had a ball catching and releasing. We had pistols and birdshot and lived mostly on fish, partridge, and fried bread.

Figure 17: Rolf and fish

We camped above the Falls Chain the first night and near Kennebis Falls the second night. There used to be a ranger cabin up there but it has been torn down since too many people liked to try to stay there. We kept going up to Kawnipi and then on to Pickerel Lake.

By then we thought we were half way to the Arctic Circle, and hadn't seen a soul until a canoe appeared in the distance on Pickerel Lake. We saw it for a few minutes, and then it disappeared into one of the bays. We wanted to connect with him, but we missed the chance. We'd heard there was a road in from the Canadian side somewhere up there, and thought maybe he came from that direction.

Anyway, we decided we'd better head south into Lake Agnes, and after trying the first two channels that were plugged up with beaver houses, we finally made it through in the third channel. We spent three days on Agnes. Agnes was a good trout lake then and still is today. I remember the first time I threw my line in with a daredevil on and wham, I thought I had a Northern, but it was a nice eight-pound lake trout. I have a picture of that fish somewhere. We ate him for dinner. I still have slides of that whole canoe trip.

You know, the Indians up here considered Northern pike to be dog food. I remember being with an Indian guide who kept saying *kenosha* every time someone landed a Northern pike. He said it with an angry tone, and we all thought he was mad at us. He finally explained that pike were considered dog food and he wanted us to catch something for dinner. I never understood it because I always liked all the fish we caught, especially fried up for dinner.

The second night on Lake Agnes we camped by Louisa Falls. We had those carbide lamps with us that miners use for light. At that time, you could get a box of carbide for a buck and we had a lot of carbide, thinking we might be out awhile. That night a bear wandered into the campsite, and we chased him out making noise with pots and pans. The next night we figured he would be back and we planned to fix him.

We'd brought along a ham for the trip, and by now, it was getting pretty ripe. We took a chunk of it, hollowed out the middle, poured some carbide into it, and put it right by the fireplace. Sure enough, that night the bear came right to the fire. They usually head to the fireplace because that's where all the odors are from all the cooking. That's also why you should always rinse out your canoe if you've been fishing. A bear will smell that fish odor and start clawing to get to it. I've seen many a canvas canoe ruined that way.

Anyway, that bear plowed into that ham and all of a sudden, he stood straight up, got real rigid, and bolted straight for the water. Of course, the water made it worse, combining with the carbide and all, so he jumped back on shore, and was spraying feces out his back end all the way to the woods. After that, we never saw him again. We had quite a laugh with that one.

From Louisa Falls, we headed south into Basswood, and we decided to look around a little. We found a little portage store, actually, it was kind of a shack, run by the Kosha brothers. They were the first people we saw, other than those in the canoe we saw on Pickerel. I often wondered who would have found us if something had happened to us. Before the war, they claimed the Canadians would patrol up there, but they must have stopped after the war because we sure didn't see anyone. Anyway, the Kosha brothers were fisherman and fished that whole area and had that little portage store where they sold a few things. In fact, they built a couple of the spillways between Birch and Carp lakes. They were interesting, so we camped nearby and stayed around there a couple of days, then headed back.

We'd been out a couple of weeks and thought we ought to be getting back. We ran the border for a while up through Knife Lake, cut over through Eddy Lake, up through Ogishkemuncie, Jasper, Alpine, and Seagull lakes and in. We were out sixteen days.

When we got back, Russ Blankenburg, who had rented us the canoe, asked if a couple of us could stay awhile to work. He said he wouldn't charge us for the canoe if we went to work for him. He was getting ready for the fishing season. One of the other guys and I said we could stay a few days. Since I didn't have any commitments at home, I figured, what the heck, so I stayed and never did go back. I've been here ever since.

I started guiding for Russ. I guided some day trips fishing on Saganaga and Seagull lakes. Then I got a lot of overnight trips. There were some Indians that guided for Russ at the time, who liked the one-day fishing trips rather than the overnight trips, so I used to take the overnight trips. That's how I got to know the country real well. I don't think I got any further south than Gunflint Lake that whole summer.

I thought I'd be heading home at the end of the season, but Andy Mayo, who owned Wildwood Resort also on Seagull Lake, was building a big place for an executive from General Motors. The guy's name was Peter Blomberg, and he was with the Buick Division of General Motors. Andy asked if I would help him build it, and I agreed and so he hired me. He gave me my own cabin and took care of my food, and I stayed the whole winter. I learned even more about the people along the Gunflint Trail since I stayed there for my first full camping season as a guide and then as a resident for the winter.

It turned out to be a timely decision to stay on and work for Andy, because that's how I got my cabin on Seagull Lake. Andy owned one-hundred sixty acres that ran across Seagull Lake for six miles, and at the northern end of that property he sold three acres to two nephews of Ted Carlson, who used to come up to the area in the summer. Andy sold it to them before World War II. Originally, they wanted to start a resort on the property. I guess that never worked out because the boys married after the war and seemed to lose interest in the resort idea. When the two nephews lost interest in coming up, the property and cabin weren't used that much.

The owners sent a letter to Andy to see if he might be able to find a buyer for the property, and I happened to be with Andy when he got the letter. After he read the letter, he asked if I might be interested in finding a buyer for the property along with the cabin. I told him he had his buyer, and it was me. I had a little money from my Coast Guard days, and I bought it with the money in savings. I thought maybe I could do some freelance guiding, or maybe even start a resort myself. Well, that's exactly what I did.

For about five years, I lived on Seagull Lake in the cabin, and did freelance guiding for a number of the lodges. It was a nice little cabin. I had an outhouse and a kitchen stove that worked with wood or propane. That was handy because if I needed to cook without starting a fire I could. Andy, Al Hedstrom, and Russ Blankenburg called the most. The owners of Wildwood Resort and Windigo Lodge called me when they needed a guide, and so did Justine Kerfoot, and I would guide for them on Gunflint Lake and some of the connecting lakes and rivers. Between the six of them, I stayed busy and had a nice variety of trips.

If I got fishing trips, they were for one day, but like I mentioned, the Indians preferred the day trips so I took the camping trips. Most of the trips were for a week or less, and usually for three days. Very few were over a week, and most of mine were small groups or families who wanted to camp and do some fishing. I did guide some small school groups, but not very often.

Figure 18: Rolf's cabin on Seagull Lake

For some reason a lot of people wanted to go to Canada, and at that time there wasn't a fee or check in. All that was needed was a Canadian fishing license. Sometimes we would head over to Northern Light like the trip you just finished, or up into the Quetico following the same trip I took that first year I was here. By going out on trips all summer, I learned the country really well. In fact, that first year, I never got down to Grand Marais all summer; I was so busy.

When I would guide for camping trips, it required more preparation. We'd go over the gear people had and take along items they didn't have to make the trip easier. Even though people expected we would cook by campfire, I always took along a camp stove that was really handy for a quick lunch preparation, or if was raining and all the wood was soaked. I did the meal planning ahead of time and all the cooking and work chores when we were out.

All the people, especially the lodge owners, bought their food in bulk. Fifty and twenty-five pound bags of flour, sugar, oatmeal, dried milk, and pancake mix, all had to be measured and packaged for each trip. Peanut butter was also bought in bulk and was packaged. Sometimes people brought fresh eggs along, so I would usually cook them early in the trip so they wouldn't spoil or get broken while we traveled.

We ate a lot of pancakes, oatmeal, and bacon for breakfast, and I would make coffee in my #10 can. People always got a kick out of my coffee maker. Noon meals were played by ear, depending on what kind of progress we were making on the trip. They tended to be lighter meals, and I usually packed a ham for sandwiches, or peanut butter. Sandwiches were simple and always popular. A sandwich and a candy bar usually went over well, and I liked to make a pot of coffee, too. Cookies and candy bars are what we had for dessert and snacks. I preferred getting out of the canoes at lunchtime, so we could stretch out and relax. Sometimes we would close our eyes, lay on the rocks, and take a five-minute nap, which I think helped people to recharge, especially if we'd been paddling and portaging hard. We always tried to catch a few fish for dinner, and I would often make potatoes to go with the fish. Potatoes were a dinner staple, and I usually packed some canned stew just in case we didn't catch any fish, or we got caught out in bad weather for an extra day.

The people who camped in those days were good people, and even though I was the guide and responsible for everything, everyone pitched in and I just became part of the group.

I'm always thrilled to hear the stories people tell after they return from a canoe trip. I feel like a teacher and may have contributed to their enjoyment of the wilderness and the outdoors. Quite simply, I always had the goal of trying to get the wilderness in people. I have always lived by the motto: *Man cannot make the wilderness, but the wilderness can make the man.*

You see, nature doesn't lie. Nature is absolute truth. Nature doesn't pretend. It's just true the way it is. It's before you as reality, and you actually become a welcomed visitor. But it won't change just because you happen to be there. The storms and the wind will come, followed by the warm sunshine. All of these things go on in the cycle whether you're there or not. When you enter the wilderness, you live in the habitat of the animals and plants around you, and you survive the same way they do. You share the water and wilderness and all it has to offer, along with the beauty and the ruggedness. Leave it alone and it takes care of itself. And it changes people. I have seen people's mental health improved just from being up here.

When I first was guiding up here, we would get executives in from Chicago who were sent by their companies. There were usually one or two guys in a group who would be grumbling about why they were up here. They would say things like, 'all you have here is water, stones, and trees.'

I'd work with them, and usually by the end of their time they would be the saddest to leave. I always wanted to get the wilderness in people so they could appreciate the simplicity and rustic true nature of things, something they would never experience in the city. You'd be surprised how often guys would bare their souls sitting in an aluminum fishing boat. I always kind of felt I may have made a difference in people's lives just by connecting them to the wilderness and listening to them. I've seen a lot of minds changed in these moments.

They'd come up to me in the end, shake my hand, tell me what a good time they had, and nod their head in disbelief, saying that they never would have believed they would drink coffee made in a can, or that they couldn't believe these stones, trees, and water could be so wonderful. Usually it worked out for them. I guess that's what's been kind of my motto up here—getting the wilderness in people, and watching their lives change as a result. Like I said earlier, man cannot make the wilderness, but the wilderness can make the man. I've seen it time and time again.

Summers are busy for all of us up here. Most of us average between eighty and one-hundred hours a week from May to October. You start early and work late. Most of the socializing with everyone that lives and works up here takes place in the wintertime."

"I'll bet it's cold up here in the winter time." Uncle Bud remarked. "How would you know if a storm was coming?"

"There have been a lot of cold winters, and sometimes a lot of snow," Rolf said. "We can't always tell when a storm is coming until it's almost on top of us. Grand Marais doesn't really have a radio station, and even if they did forecast the weather, it would be so influenced by Lake Superior, that it wouldn't really apply up here. We listen to the radio stations out of Duluth, Winnipeg, and Thunder Bay. This part of the trail sits almost in the middle of that imaginary triangle, so based on what they're forecasting, you could kind of figure what kind of weather was on the way. Sometimes Duluth reports the coldest temperature as International Falls, and we might be sitting twenty degrees colder, but we don't have a reporting station up here so they don't know what our temperature really is.

I used to have a neighbor I'd call when I lived in the Quonset over the winter. I didn't have a very good thermometer, and he was over on the mainland at Seagull Lodge. I remember one February I called to get the temperature because it sure seemed awfully cold to me. He had a new thermometer, but it only went down to twenty below. I remember him saying that it had been at least twenty below for two weeks because his thermometer hit the bottom and hadn't moved since.

We have a lot of snow, too. Sometimes two feet of snow or more will be on the ground. That's why you have to have snowshoes up here. The wind will affect how much snow will be on the lakes and on the trails. The more open the lakes the less snow there will be because the wind will blow the snow to the shorelines. The lakes that are more protected from the wind will pile up more. The trails are affected by their orientation to the wind and whether they have a lot of trees protecting them. Sometimes the drifts get really high.

They keep the Gunflint Trail plowed in the winter, so if we need to get into Grand Marais during the winter to get food or get to Duluth for medical attention, we can still travel. The school bus runs every day for the kids to go to school. At the end of the season, usually in late September and before winter sets in, we head into town and stock up on the food we buy in bulk. We try to avoid going to town in the winter. We have lots of canned goods, too, and I kept them in the lake to keep them from freezing when I was over in the cabin. That's when the labels would sometimes come off. I used wood to heat my cabin, had an outhouse, and used a cooking stove that worked with wood or propane. Having the propane stove made cooking a lot easier. I have the same setup for a stove at the Quonset where we are now. The difference, though, is that we heat with oil instead of wood, and we have a refrigerator.

Most of the men that live up here hunt, and it's not hard to shoot a couple deer during the season, which will be enough meat, along with the fish we can catch, to last all winter. I learned to ice fish as a boy, but the Indians taught me a few more tricks. Ice fishing is different from summer fishing. Deer season is at Thanksgiving time, which is when we get together with the other full-time residents and lodge owners for meals and gatherings. None of us really have time to get together in the busy season. Thanksgiving is the only time we get turkey around here, and that's always nice. When I first came up to the wilderness, there weren't any moose here. They were still further north in Canada. Occasionally, we would get moose meat given to us during hunting season, but it was actually illegal since it came from Canada. The moose are starting to migrate down this way now, so it's not such a problem to have moose meat like it once was.

We cured our meat in our icehouses, and we all have ice sheds. Putting up ice is always a big winter project that goes better with more people. We usually do that around Christmas time, when the ice is around twelve inches thick. That's about the biggest block size you can carry, and we make it another reason after Thanksgiving to get together up here.

First, we have to clear the snow off the section of lake we wanted to cut. Someone with a snowplow on a truck or a tractor would clear the area. We called this the

Figure 19: Neighbors helping each other cut ice

ice field. After the snow was cleared, we would cut the whole field that we planned to harvest. We used a gas-powered saw with a big blade, and mounted runners on it so it would slide over the ice easier. We made a cut about six inches deep going both ways, so the whole field looked like a big checkerboard of twelve-inch squares.

After the field was cut, we'd go back to each block and tap the groove cuts with a hammer and chisel which caused the ice to split all the way down to the water. It's important to keep the ice field dry, because water on ice is really slippery. We'd cut the whole field before we took out the first chunk of ice. If we were careful, the ice field would stay dry and slipping into the water became less of a hazard.

One of the easiest ways to load the ice is to build a wooden skid right up to the icehouse, which is usually located near the shoreline, then hook a line to the back of the pick-up with ice tongs tied to the other end.

When we hooked the tongs to the block of ice, the truck would just pull the ice right up the skid and into the shed. With three people, it worked well. One guy was in the ice field, one was in the icehouse, and one was in the truck. The main idea is to not fall into the water channel where the ice guide is moving the blocks to hook them up.

Figure 20: Truck pulling ice block

We did have a guy fall in the hole once, and it was pretty funny. The closest neighbor I had when I lived in the cabin over on Seagull Lake was a guy by the name of Ted Carlson. It was his nephews I bought the cabin from. He built a cabin up here on Seagull after he retired as a railroad engineer. He ran the train from St. Paul to Seattle for thirty years and said he'd had enough of train whistles and built the cabin to get away from it all. He was a friendly sort of guy and always willing to help.

One year he offered to help Andy and me put up ice for Wildwood Lodge, and he wanted to be the ice guide. The ice guide is the one who floats the ice down the water channel to the water's edge where it would get hooked on the ice tongs. It's important to be really careful not to splash water on the ice and Andy told Ted to be careful not to slip, and to stay out of the water hole.

We cleared off the ice field, cut all the groves, and just got started putting up blocks when we looked back and saw that Ted had fallen in the water in the ice channel. He'd driven his old Army truck over that day since he lived about two miles away. His old truck was rough and didn't have any heat. We all had heavy clothes on because it was about twenty-five below zero that day, so after we saw that Ted had fallen in, we headed back to help pull him out. He weighed a ton with all of his clothes being wet, and it was slippery trying to pull him out because we kept sloshing water onto the ice as we pulled and tugged. We finally did get him out.

Andy tried to talk him into coming up to the house to dry off, but I think Ted was either too embarrassed or worried because Andy's wife was there, so he took off in his truck heading for his place some two miles away. Andy and I looked at each other and I said that we had to follow him because he'd never get out of that truck. He'd be frozen stiff!

We took off in Andy's truck and followed him, and got there about the time he was trying to get out. Just as we thought, he couldn't get out because he was frozen to the seat and stiff as a board. We tugged and pulled to get him out, and had to carry and drag him in a sitting position all the way to his cabin. Once we got the kitchen stove going, we were able to thaw him out, and he was fine. Ted was a good neighbor and always willing to help, but that was quite a day for him.

When I lived in the cabin on Seagull Lake, the late fall before the lake froze and in the late spring before it thawed were the times that you might be isolated for a week or two. The mail was only delivered once a week then, and I'd pick it up on Saturday over on the mainland. In the late fall when the ice was forming, you couldn't walk on it until it was thick enough, and you couldn't canoe, so you might be isolated for a couple weeks waiting for a hard freeze. The late winter and early spring were the other times that I'd be isolated. The snow would melt first and then there would be a glaze of water on the top. Then the ice would begin to crack and it looked like a honeycomb, which meant it was too weak to walk on. Finally, the ice would break into pieces. There were times in late spring when I would hike in snowshoes to get my mail and a week later, I'd be paddling over in the canoe dodging the ice chunks.

Anyway, that's how things got started for me up here," Rolf said as he finished his storytelling.

"It's getting late. I better be heading back," he said.

"Thanks for coming out, Rolf. Come back any time and tell us some more stories. I'd love to know how you and Gail met and how you started your outfitting business," Aunt Nancy said.

We followed Rolf down to his boat, and as he left that night, he pulled out and went full throttle toward home. Campfires sure are great for storytelling and bonding, and that was exactly what we did.

While we lay in our sleeping bags after he left, we could hear him all the way home. Never once did he let up on the throttle, even through that treacherous stretch of narrows. That reminded me of what Ed had said about him knowing the bottom of the lakes as well as he knows the top. No need to slow down, if you know what you are doing.

HEADING HOME

We actually slept in a little the next morning, the day before we were leaving. I think we arose at eight o'clock. The breakfast of pancakes and bacon was great, especially the bacon. We fished that morning, and that afternoon around five o'clock, Aunt Nancy, who had been after us to go swimming to get cleaned up, came up with another plan since we were not real keen on a forty-degree bath. She decided to have Uncle Bud squirt dish soap in our hair so we would have to rinse it out. She handed Uncle Bud the soap container, and he held it in his right hand. He held his beer and cigarette in his left hand, and he sneaked up behind me and squirted soap on my hair. I let out a yell of warning as he ran for the other three in hopes of dousing them as well.

As he was squirting, the container slipped from his hand and slid down the rock beside him. When Dick saw his dad drop the container, he turned around and raced back to get it before Uncle Bud could retrieve it. Dick succeeded, grabbed the soap container, and took off after him. Uncle Bud headed straight for the canoes, thinking he could jump in, push off, and get away from Dick. As soon as he got to the canoe and climbed inside, the canoe slid sideways causing Uncle Bud to do the splits and land bottom first in the lake.

We howled at the sight of him sitting in twelve inches of water with a can of beer in one hand and his cigarette, still lit, hanging from his mouth. I am not sure how he kept from swallowing his cigarette because he was laughing so hard. He reached up, took a puff, brought his beer to his mouth, took a sip, and then kept on laughing.

"I guess you all can take a bath now," Aunt Nancy said, doubled over in laughter, too.

That was exactly what we did. We jumped in, jumped out, lathered up, and jumped back in to rinse. Although it was cold, it was actually refreshing. Since we had not bathed for six days, it felt good to clean and freshen up, and I am sure we smelled better.[14]

As I drifted off to sleep that night, I thought of what Rolf said about the wilderness changing people. I played the events of the last several days over in my mind, and I knew that I had indeed been changed.

Leaving the wilderness is always bittersweet. You are looking forward to getting back home, but you find yourself not wanting to leave. I always find myself planning my next trip as I go home. I guess it gives me something to look forward to.

We loaded up the next day and headed to the End of the Trail. We stopped to say goodbye to Cliff and Hilda first and purchased some pop, candy, and postcards. Then we dropped off the canoes and equipment to Rolf and said our goodbyes to him and Gail. He asked if we were coming up next year. We all said yes.

"How about if you take an eight-day trip next year?" he asked.

"That would be great," we all said in unison.

"I have just the trip for you," Rolf said.

When we piled in the car after packing the boat, I noticed something different about my cousins and me. In a short two weeks, we had become men instead of just greasy teenage boys. We had just proven we could handle what the elements dealt and had come out stronger. I thought about the friends I had back home and wondered if a bomb destroyed their homes or some natural disaster occurred, could they find a

---

[14]Bathing in the lake is now discouraged. Soaps are to be used away from the water, and rinsing for people and dishes is to be done where the soil can absorb, well away from the water.

way to survive? Could they withstand the truth of nature? Would they have the inner confidence to know how to live in the outdoors and make a go of it? Rolf's words kept coming back to mind. *Man cannot make the wilderness, but the wilderness can make the man.*

I guess Rolf had put the wilderness into me, and I continued that way of thinking all the way back down the Gunflint Trail. The wilderness coming right up to the edge of the road was a reminder that we were only a breakdown away from having to stay another night outdoors.

Even though I was young, I knew that America had been blessed during my lifetime with a peaceful existence. I was reminded from all the movies I watched about wars and natural disasters, though, of how fragile life can be. There was a sense that having taken this trip, I not only gained the inner confidence that I could survive, but the practical experience of having faced adversity and succeeded.

When on a canoe trip, isolated from civilization, and with only your clothes, tents, a little food and a fishing pole, you begin to realize how simple life can be. You also find out how much time is spent finding and preparing food, getting water, keeping the fire going for cooking and warmth, and just taking care of basic living needs that modern civilization provides with running water, sewage, furnaces, appliances, packaged food, and warm-dry shelters. It is a reminder of what life was like for the generations before us and for those in less-developed countries. It deepens your appreciation for what you have in a more no-nonsense, realistic way. It also makes you see how much attention we spend on other things in life when the modern conveniences are provided. Unfortunately, having more time to spend on the "other things," often complicates our lives, adds stress, and prevents us from appreciating life.

Getting the wilderness in me not only gave me the chance to enjoy God's creation, the beautiful scenery, the wildlife and animals in their own environment, the clean air and water, the starlit nights, and the dark blue skies, also it provided a break from the demands of modern living to experience a simpler, less difficult lifestyle. Never in any other place have I been able to experience that total removal from the day-to-day things of life to find these moments. I suppose getting the wilderness in you means that for all these reasons you want to experience these things again. With such a vast and varied space and so much to explore, you cannot wait to get back to recapture your first experiences and have the chance to experience new things.

The bond between my cousins and me had definitely deepened. Although we had spent many years playing and camping together at state parks, this was much more. We had worked as a team and each one of us was needed to make the trip work. If one of us had been hurt, we would all have been affected. If one of us had decided to quit, we all would have suffered. Teammates on sports teams have something in common with what we experienced, but this was much deeper, because it was not just about winning. It was about survival. We all did our jobs without being told, and took personal responsibility, something that benefitted each of us as individuals and as a team.

Although we were together for five days, there was plenty of time for quiet solitude on shore. I cherished the moments that we paddled along silently accompanied by our own thoughts, and the soft dipping of the paddle as we quietly cut the water with our strokes. I liked it when we spoke, laughed, helped each other, and sometimes sang as we paddled. I loved the sunny skies and the sound of the loons, and appreciated the dry tent when the weather turned bad. The experience of this trip has affected all of us forever. I wondered if there was really an accurate way to describe what really happened to us on this trip because it was very powerful and very real.

My thoughts were interrupted by Aunt Nancy announcing that we would be stopping to do laundry in Grand Marais and that we would have tuna sandwiches for lunch.

"After we get the laundry started and eat lunch, you boys can spend some time in Grand Marais before we get on the road," she said.

Lunch was great, and it was nice to get those stinky clothes laundered. It was also nice to wear dry socks. It seemed like mine had been wet since the first day we arrived. The cushions on the seat even felt good, too, since we had not sat on anything soft for days. Best of all, the flush toilets were great.

Our trip back to Chicago took a detour because we stopped in Door County, Wisconsin, to drop Dick at some friends. They were building a campground, and Dick agreed to work for them over the summer. We dropped him off, stayed a couple of days, and then headed to Chicago. On the way home, it was decided that Bob would come home with me and spend a few weeks in Cleveland. I told him he could caddy with me and make some good money while he was there.

We rode the Greyhound bus from Chicago to Cleveland and were finally home. We had a great summer hanging around my friends. We fixed up a bicycle for Bob, and got up early each morning to carry two loops of double bags at the golf course. The caddy master liked us because we came early and went two loops. We never had complaints from the golfers, so he treated us well. When we left the golf course at 4:30 in the afternoon with a pocket full of cash, we would race home, shower, and head out to be with our friends for the evening, usually gathering at someone's house. We fixed Bob up with a date so all of us were coupled. It was a great summer of fun and laughter, and one of my favorite summers growing up. Those were the days, when we could be out until eleven at night just having fun, no mischief. Curfews did not have to be enforced, and the worry of alcohol was non-existent. There were many evenings when we talked about the Northwoods. Bob and I were always anxious to share our stories and retelling those stories and our memories of Rolf and the wilderness strengthened the bond between us.

Man cannot make the wilderness, but the wilderness can make the man.

CHAPTER 5 - BACK TO THE WILDERNESS

The school year went quickly. Going from junior high to Cleveland Heights High School was quite a change. The winter in Cleveland was gray and snowy as usual, but playing basketball helped move it along. I hoped for an early spring so I could start caddying and saving money for the canoe trip. The money I made shoveling snow got me through the winter, but was never enough to save for larger expenses. I spent my study halls planning for the trip and figuring out how much I needed to make and save to make it all work.

I had kept a rate sheet from Rolf's outfitting business so I had a good idea what things would cost. On the back were sample menus for each meal, and that became a good template for planning this trip. Dick and I exchanged phone calls, and I sent him a menu I thought would work. I also sent a grocery list that needed coordinated with Aunt Nancy's, and included a list of accessories we needed to rent. I estimated the cost for each of us to be around thirty dollars. My clothing checklist was complete, and I spent those weekly study halls modifying and finalizing my progress against goals.

The day finally arrived when I would take the Greyhound bus to Chicago. I left early in the morning and arrived late afternoon. Ed, Dick, and Jim met me when I got off the bus, and we said our hellos grabbed my gear, and headed for their house. Bob was helping Aunt Nancy with a laundromat she managed, and was not there to meet me at the bus stop.

The trip from downtown to their house was stop and go. It seemed like rush hour in Chicago lasts until eight o'clock. When we finally arrived, it was great to see Uncle Bud, Aunt Nancy, and Bob. The drill was the same as the year before. We packed our sea bags, filled the car, had dinner, loaded the boat, and got ready to leave. The plan was to leave first thing in the morning.

While Dick, Bob, Jim, and I finished discussing our trip and began preparing for bed, Aunt Nancy had other plans. She filled two squirt guns with water and handed them to Ed, saying she thought this would be a good time to liven things up. Ed came busting into the room and emptied both guns before we could get our pants on to give chase. He headed outside, and cut his foot on the storm door while running to his car for sanctuary. While he sat smiling at us in the locked car, he asked for mercy and showed us his bleeding foot. We let him out without retaliating, however, we made sure he knew we were thinking of him the next morning. Before we left, we shut off the water to the toilet, hid the toilet paper, and put Vaseline on the toilet seat. We thought it would make his life interesting and make him think about us.

Our trip was delayed by a flat tire on the boat trailer in the afternoon, but we were able to make good time to Grand Marais. We made our final grocery stop in Grand Marais at the IGA and headed up the Gunflint Trail. Aunt Nancy called Rolf a month earlier and told him that we were coming and reserved the canoes and gear.

The trail looked the same. It was still graveled most of the way, and the wilderness stood right at the edges all the way. It looked as if the wilderness would reclaim this ground in short order if it were let go. It sure was great to be back. Making that last sharp turn by the huge boulder meant we were almost there. We pulled into the parking area and made our way up the lane to Way of the Wilderness. Uncle Bud stopped the car and we piled out, heading toward the building.

Rolf met us on the porch as we approached, and we said hello while exchanging handshakes. As we all assembled inside the outfitting building, we told Rolf we wanted to leave the next day to start the eight-day trip, so we would need our gear and canoes immediately. Rolf told us that this trip would require three maps, and he began to carefully map out the route on our maps. He marked the campsites and fishing areas and made comments as well. He also thought it was a good idea to start the trip early. He told us that if we were delayed because of wind and weather, we had a little leeway to get back in time.

Rolf also mentioned that there was a waterfall called Louisa Falls at the bottom of Agnes Lake. It starts at the top, comes down to a bowl about nine feet across, and then plunges down another seventy-five feet. He added that it was calm enough to swim in, although it might be a little cold. He encouraged us to check it out.

"Remember not to get discouraged on the first portage," he said. "That Silver Falls Portage is pretty, but it's tough, especially if it's raining. Once you get past that one, the others are a little easier."

Aunt Nancy interrupted Rolf at one point and asked if she could say hello to Gail and their new baby. Sally was born in January, and Aunt Nancy was anxious to visit Gail and see how she was doing. Rolf told her that Gail was at the house, and Aunt Nancy grabbed her purse and pictures and headed out the door to find Gail.

Rolf gathered our gear and met us down at the launch for Saganaga Lake. We stopped at Cliff and Hilda's to say hello and grabbed bottles of grape pop and candy bars. We got to the launch as Rolf was taking the first canoe off the top of his station wagon. Aunt Nancy returned from her visit with Gail, and we scurried around packing the boat and canoes, and prepared to head out to our campsite. As we were leaving, Rolf told us that there was another nice campsite to the east we might like if Clark Island was occupied. Aunt Nancy seemed a little annoyed that someone might be using her favorite campsite.

"Good luck!" Rolf said, as he stood on shore waving.

We made our way out through the narrows and into the southern portion of the lake toward Clark Island. As we got closer to the opening, we noticed a tent on the island. The site was occupied, so we bore east and headed over to the area Rolf had suggested. That site was available, so we pulled in to have a look, and Dick jumped out of the boat and scurried up the hill to have a look around.

"It's not as nice as Clark Island, but I think it looks pretty good," he said, heading back toward the boat. "There's room for four or five tents and the dining fly. I think it will do just fine."

The routine was the same as the year before. Aunt Nancy was barking orders and we were scurrying around to set up camp. We started with the dining fly and worked until we had the tents up and all of the items in place. Uncle Bud started the fire, and the smell of burning hardwood was a pleasant scent to all of us. The bugs were busy, of course, so after a much-enjoyed dinner, we retired for the evening.

It is always a little hard to fall asleep the first night in the woods. The sounds and environment seemed new and unfamiliar again, strangely enough. But after a little conversation with Dick and silent prayers, I fell asleep.

THE EIGHT-DAY TRIP

We awoke to the fresh smell of bacon and coffee. Aunt Nancy was cooking eggs and bacon and told us it would be ready by the time we had the tents down and were finished packing. Breakfast was delicious, and we hustled to clean up and finish packing for the trip that we had been planning all year. We loaded everything into the Duluth packs we rented from Rolf and carefully filled the canoes, balanced the weight, and hooked the shoulder straps of the packs over the shoulder pads of the carrying yokes.

The skies overhead had turned gray after a hazy start. The good news was that the winds were fairly light as we pushed off to begin our journey northward to Canadian Customs. We were optimistic as we paddled along.

It took a little while at Customs. Dick had to get a fishing license again, and we had to review our trip and pay for the nights that we would be in the Canadian Quetico Forest. The cans of pop and Canadian candy were good and we took them along to finish on the water.

As we approached our canoes, Dick studied the sky, wondering whether or not there might be rain. The skies were certainly gray, but there really was not much wind. Bob and Jim decided to put their ponchos on now, rather than try to do it while on the water. We shoved off with Dick in the stern of our canoe and Bob in the stern of his and Jim's canoe.

The paddle across the northern section of Lake Saganaga is about six miles and wide open. The islands are located in the southern half of the lake. Not long after we departed, the wind began to pick up, and it seemed to build with greater intensity every fifteen minutes or so. Unfortunately, it was blowing right out of the west into our faces. The waves were also building and after an hour into our paddle across, the open expanse of water was beginning to white cap and the waves were rolling about a foot and a half. Chilling rain began to fall, at first as a few sprinkles and eventually a steady downpour. Paddling became more difficult because the winds were beginning to gust, and the waves continued to build.

As we continued, I remember asking Dick if he thought we should turn around and head back to Customs or keep going toward a small island we could see in the distance. He motioned and pointed to keep going straight ahead. Bob and Jim were about fifty feet away, and they seemed content with continuing on as well, so we did.

Although the island was in sight, we made very slow forward progress. The wind became stiffer, blowing around twenty knots. The waves were slapping the canoe and the rogue ones splashed over the top of the gunwales. With the now-driving rain, the canoes were becoming heavier and harder to paddle. The extra weight was actually a good thing because it kept us more stable and less bouncy on the water. It seemed like every stroke required all we had just to move forward. I kept paddling hard, switching sides quickly to keep the momentum. My shoulders burned from the strokes. Dick had his hands full trying to steer as the waves bounced us off course in both directions.

After a steady hour of relentless paddling in the worst part of the storm, we finally came into the lee of the only island in this part of the lake. The canoes had water sloshing around in them, but the wave action was much more subdued here. We were able to find a place on the island, and pulled the canoes up so we could get out of the weather.

We were soaked and cold so Dick started working on a fire. Jim and I scouted around the island only to conclude that there was no way we could camp there. There was not one level spot on the whole island. The little fire was not much, but it helped warm our hands and arms a bit. We sat tight for about an hour and a half, and finally the storm weakened. Luckily, this was not a thunderstorm like most of them become. The rain reduced to a sprinkle and the wind backed down to about ten knots.

I decided to start bailing the canoes, and while trying to punch down the lid of the pop can with my hunting knife, I managed to put a nice gash in my finger. The last thing we needed was to try to find the first aid kit in the rain, so I tied my handkerchief around it to stop the bleeding. It took awhile to bail the canoes, and by the time I finished, Dick and Bob had arrived with some saplings to use for lashing the canoes.

We remembered what Rolf told us about lashing the canoes. The wave action had reduced and the winds were diminished. The conditions seemed okay for lashing now, although they certainly had not been when we crossed the open water of Saganaga Lake. We thought we could at least make it to the ranger station in Cache Bay. I was glad we were loaded, because the canoes were much more stable than when empty.

We finished lashing the canoes and pushed off to head across the last portion of Saganaga Lake toward the ranger station. Paddling was a little easier since the canoes had less water in them. The wind was at about ten knots, with the waves running eight to ten inches, and the rain was more of a drizzle. We were all still wet, uncomfortable, and tired from the first six-hour ordeal.

We finally had the ranger station in our sights, and this gave us a little extra energy as we steadily stroked our way toward our destination. Tired and now really miserable, we finally pulled into the dock at the ranger station. We arrived just in time because as we tied up, the rain began to come down hard again.

As we headed up the path toward the ranger station, the ranger encouraged us to hurry out of the rain. He told us that we had him in a panic when he saw us coming across the water. Not more than a week before, two men drowned when the poles they used to lash their canoes snapped and buckled their canoes under. He told us lashing canoes could be dangerous because of capsizing.

"Thank goodness you made it," he said. "What in the world were you doing out there?" We explained what had happened that day and told him that we would have stayed on the island we had fetched, but could not possibly have camped on it. That was why we continued on with our goal to get here. We told him we had been aware of the dangers of lashing canoes, and that it was dangerous in wave water, even though we thought that the conditions were favorable this time for lashing. Between what Rolf told us, and now this man, that was the last time we ever lashed canoes. The ranger insisted we stay the night in his boathouse and told us we could cook our dinner in his cabin.

We followed him down to the dock and paddled our canoes in as he opened the door to the boathouse. There was not much room in the building because the ranger's boat was in there, too, but there was plenty of space around the boat and we unloaded our gear.

The ranger noticed the bear-paws on the canoes and asked if Rolf had outfitted us. We said yes. He knew Rolf from the time he had been there. I got the feeling that the fact that we were outfitted by Rolf may have communicated to the ranger we were more prepared than other parties. He did not say that, but it was conveyed in the way he made his statement and asked his question. He treated us as though we were perfectly capable, rather than rookies who needed tender loving care and coaching.

Between the rain and the water in the canoes, half of our gear was wet, including most of the bread on the top of our packs. We spread the gear out as much as we could, and hung up everything else. It looked like a Chinese laundry in there and we used all the space we could. We all changed into dry clothes, grabbed our hamburger and damp buns, and headed for the cabin to cook dinner.

The cabin was warm, unlike the chilly damp boathouse, and felt very cozy as we cooked our dinner. The smell of fresh hamburgers filled the cabin and all of us gobbled down our dinners since we had never really had lunch. The food not only satisfied our hunger, but also warmed us. We spent a little time visiting with the ranger after we cleaned up our dishes, trying to stay in the warm cabin as long as possible.

We talked about our previous trips and experiences while in the wilderness. The ranger added his comments and talked about his job and the people who came through.

When we headed down the path back to the boathouse, we were practically running because the rain was coming down again. At least we would be dry inside the boathouse tonight. We spread our sleeping bags out on the docks and fell asleep listening to the rain on the roof and the water lapping the pilings of the boathouse. The rain continued throughout most of the night, stopping at around four o'clock in the morning. The wind got stronger during the night, and when we got up the next morning, it was blowing steady at fifteen knots with gusts to twenty.

We went up to the cabin to cook our breakfast of bacon and eggs and visited some more with the ranger, finalizing our plans. We told him that we had hung up our gear, and it had not dried much, but that if we got over the Silver Falls Portage, we could set up camp and get things dried in the breeze. The ranger agreed that our plan was good, but with the wind and chop on the bay, he offered to tow us across to the portage.

We went back to the boathouse, packed our gear, and paddled out to the dock. That would make it easier to hook up for the tow. The ranger met us there and told us how he wanted to arrange the canoes for towing. They were still lashed together, so we tied a line to both bows and a single towline to the middle forming a triangular sling. We left enough distance on the line to keep the canoes behind the wake of the boat. The ranger went slowly so there was not too much sway in the lines, and to keep the eight-to ten-inch waves from splashing water in the canoes. Even though it was still windy and there was a chop on Cache Bay, the sky was deep blue with puffy clouds racing through the windy sky. The sun was beautiful, but the wind made it a chilly and snappy morning on the water. It seemed like the rain had let up for the day.

The ranger towed us down the narrows past Indian Grave Island to the Silver Falls Portage, and cast us off near shore. He waved good-bye and smiled as he headed back. We quickly unlashed the canoes, and started our first trip over the portage. It was hard, especially with the packs full for eight days and still damp. We got some great pictures of the falls, which is gorgeous as it plunges down eighty feet and over three-hundred yards.

On the other side, we loaded again and started looking for a campsite. We found our site not far from the portage, set up camp, and began hanging clotheslines from every tree we could find. Some of our bread had been ruined in the storm so we threw it out for the birds. We were able to salvage some of the loaves, and thought the peanut butter would help hold the smashed pieces together. Once we had things in order, we decided to do some fishing.

Luckily, the weather cooperated and we were able to get things dried for the most part. I say for the most part, because it never seems everything is completely dry, especially your shoes and socks. We caught some fish below the falls, and ate like horses that night. When we turned in for the evening, we knew we were already a day behind and that to stay on our eight-day schedule, we would have to make up for our lost day. Rolf always gave people an extra day or two before he began to worry. Being weather-bound is a common occurrence on trips. What we did not know is what other surprises lay ahead.

We packed up in the morning and headed northeast to Dead Man's Portage[15] with our goal to get into the Falls Chain. The Falls Chain is a series of falls that lead north into

---

[15]Dead Man's Portage was named that after they found a dead man at the end of the portage.

Kawnipi Lake from Sagnagons, and the lakes connect to the north. Rolf spent a lot of time telling us to stay close to shore as we approached each portage.

"You must be careful up in those falls portages," he said. "The current is strong, even though you can't really see it. Over the years, someone has been killed at every one of those falls. They get out in the middle and are swept over. You'll be fine if you stay near the shoreline. Just be careful."

Rolf's words echoed in my mind as we headed down the channel for the first portage. This advice hit home with Bob later in life when he and a friend came through the Falls Chain in the 1970s. On their way out from a trip, they encountered a battered canoe and gear littering the shore below one of the falls. The word *HELP* had been written with a charred and blackened log on the face of the rocks. No one was around, but they found out later that two guys in a party of four had been swept over the white water, and the rescue team found their bodies pinned in the rocks in the middle of the white water.

Figure 21: Part of Falls Chain

We made three of the portages, and camped at the bottom of the fourth. When we pulled into our campsite, we noticed that one of the canoes was leaking along the seam at the bow. It had a slight separation. We thought it may have stressed in the storm, and we decided to tap a pine tree, let the sap run out, and see if we could force it down into the seam. We thought we could even add some pine needles to help hold it together.

We tapped the tree, and then went fishing below the falls. We fished from the shore, climbing on the rocks while waiting for the pine tree we tapped to emit enough sap for the canoe repair. The fishing was great, and we caught walleye left and right. They looked good in the frying pan and they tasted even better. The next morning, Dick and I caught a few for breakfast and fried them as we finished packing for the day ahead. We added more tree sap to the canoe, which had been repaired the night before. It seemed to have sealed the seam and hardened. We added a little more for good measure and headed out.

Our trip up Kawnipi was uneventful with just a few showers here and there. Our repair fixed the canoe leak, but we added sap every evening since the leak persisted, although much less. The wiggling of the canoe in the water and the landing and leaving portages tended to stress the seam and cause it to leak as the day wore on. In addition to having a goal of using only one match per fire start, we had a goal of having a whole day of no water in the canoe because of our newly found repair skills.[16] We managed to achieve the first goal, but only managed a little success on the leak.

We stayed on the northern section of Agnes Lake, and with a favorable north wind the following day, sailed down Agnes and stayed at the southern site beneath Louisa Falls. Sailing in a canoe is accomplished easily when the wind is from behind. The front paddler ties the two top corners of his square rain poncho to two five-foot sticks. Holding the sticks in the air, leaning back, and holding the bottom corners with his feet presents a nice square sail for the wind to push. The rear man only has to rudder the boat to keep it on course. We moved down the lake quickly with occasional changes in the front and back to relieve the arms of the front paddler.

---

[16]These were the days before duct tape had become the first defense and repair for almost everything.

When we reached our site, Dick ran ahead and motioned for us to get up to where he was at the top of Louisa Falls. He could not yell, and instead was waving and put his hands wide-open, thumbs to his head, trying to imitate moose antlers. He had seen a moose and wanted us to come up and have a look. The moose was gone by the time we got up there, but we were able to see the bathtub formation Rolf told us about when he told us about Louisa Falls. It was beautiful as the water plunged down out of that formation to the rocks below, but cold, and it was unanimous that we were not swimming today.

We had a good laugh, though, when Dick grabbed the toilet paper and headed up the steep hill to do some business. After a few minutes, we heard him call for help because he lost his paper. When we turned around, we saw the roll of toilet paper unrolling all the way down the hill. None of us were planning to help him, and by now, we were laughing hysterically at the scene.

Figure 22: Louisa Falls

By the time Dick finished, we settled down, but Dick had the last laugh when he placed the now loosely rolled paper, pine needles, sap, dirt, and all, back into the pack.

"You guys can use this roll," he said. "Just pick off the pine needles."

When we left the next morning and headed south to the two long portages toward Prairie Portage and the American side, we heard a commotion on the water. As we got closer, we saw two canoes in the water paddling behind a bull moose that was swimming in the water. They were Boy Scouts who were part of a larger group. We shook our heads in dismay, wondering if they had any idea how dangerous a moose can be. One flick of those antlers would spin a canoe right over. Moose are good swimmers and powerful. If the scouts dared follow the moose on shore, they would be in more danger. A charging bull moose is no laughing matter.

Rolf had explained the difference between moose and deer to us. He said the Indians would sometimes chase deer to the water, because, although a deer could swim, it did not have much endurance. Once in the water, the deer would get tired and actually drown, and the Indians could pull him back to shore, and have deer that had not been injured from hunting. They even said that a deer could be run to exhaustion, enabling them to be captured and killed. A moose on the other hand, can run and swim for miles. Rolf said he had a bull moose jump out in front of his car once while driving up the Gunflint Trail. He said that moose stayed in front of him for almost twenty miles, trotting along, before he finally cut back into the woods. Rolf swears that you cannot wear down a moose.

There was a sense of sadness that our first encounter with people was this scene. We paddled on and kept moving. It was nice to get to the U.S.-Canadian Customs and re-enter the U.S. We had candy bars and pop while we were there, but we knew we were still a day behind and had a long push ahead of us.

The wind turned to the southwest the following day, and we decided to break out the sails again. We would be heading northeast almost all the way back and with this stiff breeze we could make up for our lost day. We sailed all day up through the Knife Chain and into Ottertrack. What a day! We made up for our lost day, and if everything else went according to plan, we would be back to base camp tomorrow.

Figure 23: Paddling to Customs

We woke up that morning to a sunny sky with a breeze that was lighter than the previous day. We made the Monument Portage, through which the international border runs. Small cone-like structures about three feet tall had been erected marking the borderline. This was the only time we saw a manmade structure on any portage.

The long paddle across Sag was not nearly as eventful as our first day out. The wind was mainly at our side as we made our way to the site where Uncle Bud and Aunt Nancy were waiting. When we pulled in, Uncle Bud and Aunt Nancy greeted us at the shoreline. We scrambled out of the canoes and began a non-stop conversation about our trip, and what had been going on in our lives in the last eight days.

Early in the conversation, we noticed that there were extra lines tied to the dining fly. Uncle Bud told us that the same storm that wreaked havoc on us the first day out practically blew the tent down and collapsed the dining fly. Aunt Nancy laughed as she told us that when the wind started to blow that day, Uncle Bud was racing around tying everything down using every granny knot he knew. She told us there were so many extra lines everywhere they could hardly walk, and the reason they had not moved the campsite was because Uncle Bud did not want to untie everything.

Aunt Nancy then called me over and asked me to go for a little walk with her. We walked down by the water's edge and she asked if I would like to stay up here and work for Rolf for the summer. She told me that she asked Rolf if Jim and I could stay and help him this summer, and that she had already cleared it with my sister Diane if I would like to stay. Rolf told her that we could clean gear, stencil, and generally help him with whatever he needed in the business. At the end of the summer, I could go home on the bus. I was thrilled and said I would be glad to stay.

We spent the night at the base camp and packed up the following morning to head home. That bittersweet feeling was lessened this time because I knew I was staying. I could not wait to see Rolf and tell him about our trip. He always took such a genuine interest in hearing about the adventures of his customers. We landed on the dock and began unloading and repacking the boat and car for travel. When things were finished, we headed to Rolf's building, and he greeted us with a big smile.

"You made it!" he said.

"We had a few problems along the way," Dick said with a smile.

"I heard about the storm on Sag, the soaked gear and food, and the leaky canoe," Rolf said.

We asked how he knew.

"Moccasin telegraph," he said. "It's pretty reliable up here, you know."

We continued with our stories and Rolf asked questions about our trip. He listened intently, anxious to hear everything. As the conversation wound down, Rolf directed

Jim and me to take our bags upstairs to the loft of the outfitting building, and told us he would get us organized up there when everyone left.

As we said our goodbyes to Uncle Bud, Aunt Nancy, Dick, and Bob, I had the same feeling that I had at the end of our last trip. Once again, we had grown as men. We had faced hardship of monumental proportions and had endured. I wondered how many would have actually come through what we did and not be inclined to turn back. Turning back was never considered because that would have been failure. The only thing we knew was to persevere and to finish the task. The only story we wanted to tell was the story of how we made the trip, not how we quit.

I still believe that today. The commitment is to finish what you start, no matter what the difficulties. Period. The only story worth telling is the one of endurance and perseverance, and the one that recognizes the hand of God at work in our lives. This experience gave me an inner strength that would last forever. It also seemed that the wilderness was a little deeper in each of us, and that our manhood was indeed being shaped by these experiences. I guess you could say the wilderness was creating men.

We watched Uncle Bob's sedan pull away, boat in tow, and Jim and I walked back toward the outfitting building. This would be home for the summer, and I wondered what would be in store.

CHAPTER 6 - SUMMER WITH ROLF AND GAIL

When Jim and I reached the outfitting building, Rolf was hanging up the phone. He motioned to follow him to the back section of the outfitting building. We followed him up the ladder leading to the loft overhead. There were bunks at each end of the building, centered in the room with the slanting roof as a ceiling. We each chose a cot at opposite ends and dropped our gear. Rolf showed us how to open the windows on either side in case it got stuffy, and told us to take it easy for the rest of the day. He said Gail usually had dinner ready at 5:30, so we should be ready for that. After he left, I decided to take a nap, and Jim went to Cliff and Hilda's for a bottle of pop.

After resting awhile, I went downstairs to see what was happening. I had heard Rolf on the phone with someone, but it was quiet now. When I got to the office, he was gone, but I had heard the door slam so figured he must still be outside.

As I looked around the room, I was drawn to the wilderness maps on the wall. These were the same ones I studied when I came to Rolf's outfitting building the first time. I stood there, amazed at the enormous expanse of wilderness, and wondered if I would become familiar with some small part of it. When I traced the two trips we took, they seemed so small and minor in comparison to the whole area. Yet I remembered that we covered a lot of water and ground. We saw some beautiful country, and it seemed that we saw a lot of it. In contrast, though, it was just a small amount of this vast area.

I kept thinking about Rolf and how much of that wilderness he had seen. I was still amazed at how much he remembered about all of these lakes and portages, given how expansive the area was. I must have been standing there for at least twenty minutes, when I heard Jim and Rolf's voices as they came up the steps into the building. Rolf asked me if I was planning my next trip, and I think I may have answered with a yes and no.

I asked Rolf if he had been on all of the lakes and he told me he had not. He said there were over 1,500 lakes between the Minnesota and Canadian sections, with over 2,000 miles of canoe routes. He traveled most of the main routes, but unless you set up a base camp for a few days and take side trips off the main routes, you would not get into some of the more remote areas. He told me that one of the reasons he always listened carefully to people when they talked about their trips was to learn something about an area where he may not have been yet. Even when you have not been to a certain lake or area, you could still relay any information you heard from someone else. Listening to experienced canoeists helped his knowledge, beyond his own firsthand experience. He said information from reliable sources was just as good as first hand in some cases. Much of his guiding, although extremely extensive, occurred from 1946 until 1956. After he started the outfitting business, he was not able to get out as much, so he relied heavily on the information of others. I smiled and asked him if that was like moccasin telegraph.

"Kind of, I guess maybe more like moccasin tales," he said with a chuckle.

With that, we headed up to the house for our first meal with the family. Meal times, especially supper, were eaten outside on the picnic table. With their kids, Jim, and me, there was much more room outside. Clean up was easier, too, because the critters got the crumbs, and the ground soaked up the spills. There was no getting away from the bugs as we ate, though.

Meals were simple. We usually had cereal in the morning in the outfitting building. Lunches were typically peanut butter and jelly sandwiches, although sometimes they were bologna or cheese. Dinners were meat or fish and potatoes, noodles, and a vegetable. It seemed like the kids spilled more than they swallowed, but it was always fun to be around them. Gail certainly had her hands full, especially in that small Quonset hut. The kids played outside as much as possible, and wandered up to the outfitting building occasionally to say hello. Most of the time they were busy being kids and doing the things kids do. Once, while eating dinner, I asked Gail about school and how they handled it in the winter.

Gail told me that the school was in Grand Marais, and that the kids had to get up early to catch the bus. They ride halfway down the trail, then change buses to go the rest of the way. It was an hour-long ride, and it made quite a long day for them. She told me the state keeps the Gunflint Trail plowed all winter so they can get up and down it all year. She explained that when there was a bad storm, everybody helps each other. If the schedule changes or the bus is late because of weather, they make phone calls and keep people updated. She said they all just work together.

It would still be light outside, so Jim and I would either fish, or hang out after supper each night until it was dark and time to go to bed. Friday nights were movie nights. The Parks Department continued to show movies on the side of Rolf's outfitting building, so that was always something we looked forward to. The crowd was always pretty good, because people camping at the campground attended in addition to the residents around the End of the Trail. It was typical to see crowds of twenty to thirty people on movie night.

It was slow that summer, so Jim and I got all of the canoes cleaned and the equipment re-stenciled with Rolf's bear paw. Rolf made the original stencil on a couple of his packs by placing a dead bear's paw on an inkpad and stamping his equipment. He adopted the bear paw as his logo when he went into business, and had an artist develop the artwork. The stencil we used on the canoes said *Way of the Wilderness*, in addition to the bear paw. It gave some additional opportunity for advertisement.

Because business was slow, Jim went home after about ten days, because there just was not enough work to keep us both busy. One of the larger projects I did right after he left, was to cut down the brush along the south side of the lane leading to the outfitting building. It made it much easier to see Rolf's place from the road. We piled up the brush, poured gas in a can, put it in the brush, and threw a match on it. Gas does not explode in an open vessel. Rolf left me there for the day to watch it and came around once in awhile to check on things. He told me to let the fire burn over to the edges, so we would not have to cut for a year. He also said not to worry about the roadside because it was gravel, but when it got near the edge that we had not cut, I would need to stomp it out. He called this a controlled burn, and said it was safe when there was no wind blowing because the fire would only go where you let it.

I was impressed with his knowledge as always, and learned how to start a fire and conduct a controlled burn that summer. I smelled like smoke at the end of that day, but it was interesting to see how to burn a half acre. In addition to learning how to chase bears out of the garbage cans, I wondered what else I might learn.

LEARNING MORE

Most mornings were cool when we awoke, so Rolf lit a fire in the stove he had in the outfitting building to take away the chill. One morning he told me that we were going to go over and work on some cabins he had been building after lunch each day. He had been hired by Al Hedstrom from the End of the Trail Lodge to build more cabins, and he had not completed them over the winter. He needed to finish the siding and some other things, so we would work that into our schedule. He said that siding would be much easier to hang with two of us.

I heard the door to the outfitting building open as we finished talking, and someone came in. It was a man and his son. Rolf greeted them and I listened to their conversation. The man wanted to take a canoe trip with his son, and Rolf was feeling him out as to his experience. He determined that they were rookies and suggested the three-day loop through Seagull, Alpine, Red Rock and Saganaga lakes. He detailed everything, just like he always did with us, mapped out the route, and planned the trip for them. They had no gear except fishing poles, so Rolf was going to do a complete outfitting.

Rolf told the man to go up to Cliff and Hilda's store at the top of the hill to get a fishing license as he got things ready for them. He said it would take him about an hour to finish, and that he could bring their clothing and fishing equipment after he got the license. He would help them finish packing and send them on their way.

I asked Rolf if I could help him and he politely said no. He told me I was welcome to watch and could help him load when he was finished, but he had a system and he had to concentrate on every detail. One forgotten item, such as matches or a can opener, could ruin the whole trip. He was very fussy about how each pack was organized and packed. He also asked that I not interrupt if possible, although I did get the tent, sleeping bags and Duluth packs pulled out.

Rolf meticulously prepared the food and cooking items, checking everything against a printed list. He showed me how he packed everything and explained why he did things the way he did. Years of practical experience guiding and traveling the country were the reasons things were done the way they were. He wanted everyone to have a good experience.

When the man and his son returned, Rolf went through the menu and checklist, and he told them where each item was located in each pack. We packed their clothes in the last pack, and loaded everything into the station wagon. We lifted the canoe on top and headed over to the Seagull launch, where we saw them off.

"I'll be surprised if they stay out the whole trip," Rolf said as they left.

"Why would you say that?" I asked with a look of surprise on my face.

"You can just kind of tell. We'll see," he said. "I could be wrong."

That afternoon we loaded up the tools and headed to the cabins Rolf had started in the winter. The first one we pulled up to was one that needed the siding finished. They were very nice cabins, and I asked how he was able to build them in the winter.

He told me it was definitely cold then, but once you get the shell up it was a little better. He used the wood stove he had in the outfitting building to keep warm. He said he sets up the stove and gets it going, then throws scraps in it to keep the fire going. You can still put in a good day even when it is twenty below zero, but you have to stop once in a while to get warm when it gets that cold. We got a lot done that afternoon. We worked on the siding and Rolf even let me make some cuts with the power saw.

We made several trips to work on the cabins that summer, and in the process, I learned how buildings were built. Rolf showed me how to use cleats for putting up clapboard siding, and explained the basic construction of a building. He talked about twenty-four and sixteen-inch centers, headers, and corner posts, and he showed me examples as we worked. I would come back at night and write everything down and draw sketches so I would not forget. When I returned home that summer, I bought some balsa wood and built a model of a building using my notes and sketches. This was knowledge I could use for a lifetime, and I have built many buildings and done a lot of remodeling using what I learned helping Rolf that summer.

As we returned to the outfitting building and got out of the car, we could hear Gail laughing while she talked to a rugged-looking boisterous fellow over by the picnic table. They seemed to be having a good time.

Figure 24: Ollor, Gail, and kids roasting hotdogs

"Sounds like Ollor's here," Rolf said. Ollor is a friend of ours who stops in once in awhile to say hello. He's been around the wilderness for years. I'll tell you more about him later."

As we walked up to Gail and Ollor, Gail was pleading with him to stay and have supper. He was a little reluctant at first, but finally agreed. As we got closer, Rolf greeted him with a big smile and introduced me. Gail motioned for us to sit down so we found places in between the kids to sit.

Ollor was quite a character. He was half Cherokee and played the banjo. He was a storyteller, and he, Rolf, and Gail talked about people and places they had in common, and caught up since the last time he was around. It was fun to listen to even though I did not know the people. He stayed for nearly two hours that day, and it was a fun visit. When he left, Rolf told me Ollor's story.

Ollor's real name was Rollo Stevens. He goes by Ollor Snevets, which is Rollo Stevens spelled backwards. His past was a little bit of a puzzle, and he does not talk about it, but he came from Michigan during the logging days. Al Hedstrom met him in Duluth around 1952, and he wanted to do some guiding. He hauled gravel and worked into guiding down on Poplar Lake. He was in his fifties at the time. The Army Corps of Engineers hired him and he set out with two other people to finish mapping the wilderness.

They took a square stern canoe and started out on Basswood Lake, working their way up Knife Lake along the border, and back to Poplar and Gunflint lakes.

Ollor would drop in and out of sight when he was not guiding, but when he needed equipment, he rented it from Rolf. He also explained that Ollor could go through a forest fast when he was logging. He lived in a tent in the winter, and he had a smaller tent inside the larger tent to keep off the snow. He would cut down fir trees, and place them with the trunks facing up over the ridge of the tent, alternating with a tree over the right side followed by a tree pointing downward on the left side. By alternating sides like this, they would interlock, trunk to trunk. In between the first rows of trees he would add smaller ones to give a fuller cover to the tents. Because the trees were upside down, the limbs would actually help shed the rain and snow away from the tent. To keep warm, Ollor would dig a hole to put his small wood stove at the head of his tent. He would dig another hole, and run the stovepipe under the ground before it made the bend to go up on the outside of the tent. He would line the trench that contained the stovepipe with rocks, so the heat was absorbed by the rocks that caused the heat to radiate inside the tent. This kept his cot warm at night, and kept the tent temperature comfortable. Rolf said that he and Gail snowshoed over to see him one winter day and found that the tent was quite comfortable.

Rolf told me about another enjoyable fellow named Arnold Winkler, who was from Sweden and came there in the forties. Arnold snowshoed one-hundred miles from Canada somewhere west of Thunder Bay one year just to inquire about guiding. He had six dogs and he claimed they kept him warm when he traveled. A "four-dog night" meant it was not very cold, but a five-dog or six-dog night meant it was really cold. Unlike Ollor's stovepipe and tent rig, Arnold used his dogs to keep warm in the wintertime. Gail used to save food for Arnold's dogs when he lived on a small island on Saganaga Lake. Rolf said he was quite a clown and kept people happy.

Ollor was quite a guy, and he really liked Gail. They always got along, and Rolf told me years later that Gail visited Ollor in the hospital when he was sick and was at his bedside when he died.

At some point, I asked Rolf about the sign we always passed that said "Wilderness Canoe Base" over by Seagull Lake. I had noticed it several times when going up and down the Gunflint Trail on the way to and from the cabins we had been working on.

Rolf told me that it was the Lutheran Wilderness Ministry. Ham Muus started it in 1957. Ham brought inner city kids up to experience the wilderness. He said he came over to his cabin on Seagull Lake when he lived there, and asked if he could help them with their December ice harvest their first year up here. They acquired Fishhook Island on Seagull Lake, and since it did not have any electricity, they thought they

Figure 25: Dogs used for warmth in the wilderness

might need ice for their icehouse. Rolf said he saved the day for them. Ham and his friend, Bob Evans, did not have any experience, and the biggest concern with the inner city kids they had with them was about keeping warm. Rolf helped them lay out the ice field and start the cutting, and generally get organized, and some other fellows built an icehouse out of trees that had blown down.

Rolf was a little worried at first about them sending all those city kids out on canoe trips and how they might treat the wilderness, but he was impressed with how they handled themselves. They treated the wilderness pretty well, and Rolf said they have sent many people out over the years.

In the winter of 1959-1960, Rolf said he helped them move cabins out to their island. A bunch of volunteers helped dismantle some cabins over in Ely, and a man named Ed Thoreau and his crew hauled them up to the End of the Trail. When the ice formed, they were able to make a roadway from Blankenship's Landing to the islands. The only way to transport big items like logs and gravel is in the winter when things freeze. Rolf would go by snowshoe from his cabin in sub-zero temperatures to help. The logs were numbered, and they had to get them over the ice before the spring thaw. He said they made log chutes up the cliffs to haul the logs and the gravel for the concrete. He thought they might have used the same principles as the pyramids. Ham's wife, Pearl did all the cooking, and they always had a group of volunteers who helped them get the job done in time.

Rolf explained that it was actually an outreach ministry. They bring inner city kids up for canoe trips and wilderness experiences. He told me Ham comes to visit once in a while, and they all go to Sunday service out there when they are able. Rolfs twins, Stuart and Stanton, were baptized in the Quonset house by Ham Muus.

Rolf said they had worked hard out there and had become good neighbors. He also said Ham thanks him often for teaching him about the country and the wilderness way. They got along pretty well, even though they are from different backgrounds.

Gail had been busy getting the kids ready for bed while Rolf and I were chatting. I always headed in early to read and relax, and it enabled Rolf and Gail to have time with the kids at bedtime. They did not need me around all of the time.

The next morning, I got up and ate breakfast. Rolf came in and saw me studying the Hamm's beer advertisement, the one with Rolf in the red Old Town canoe. Rolf was the early star in those shots, although later he helped the company with photos of Earl Hammond and his three-hundred fifty pound Kodiak bear named Sasha in their later promotions. Those were ads created on Seagull and Saganaga lakes, and Rolf was the guide.

Rolf also told me Earl's family has provided most of the animals used in advertisements over the years. They have a big place back east in Pennsylvania. They are the ones who provided the stag in the Hartford commercials. They also provide the reindeer every year for the Macy's Thanksgiving Day Parade, as well as this Kodiak and other animals.

Rolf said it took three days before he could get close to that bear. Bears have to warm up to a person slowly. He said you had to put your hand out and let the bear sort of "mouth" your hand. It was really scary the first time, and he always remembered that they were wild animals and never really fully tamed.

Rolf said he had a lot of fun doing those ads. People in town could not understand why they were buying up so much dog food until they were told they were feeding a three-hundred fifty pound Kodiak bear.

Since we had been talking about advertising, I asked Rolf how he advertised. He said they had done a little in outdoor magazines. A man named Bob Cary from Chicago helped him get started. He was the editor of the *Prairie Sportsman*, and became the outdoor editor for the *Chicago Daily News* in 1958.

Figure 26: Hamm's Beer bear nuzzling Rolf's hand

In the summer of 1958, Bob wanted to set a record canoeing with a motor from International Falls to the Pigeon River near Grand Portage. He planned it for the middle of June, so he could run day and night during the full moon. He figured the water would be up then as well. Rolf got a little publicity from that, and Bob ran an ad for Way of the Wilderness because Rolf helped him.

Bob wrote to Rolf the following summer, and asked him to stash two cans of gas at Monument Portage so they would be there when he made the run from Fowl Lake to Fort Francis.[17] He was a bit upset with Rolf because he could only find one can. Rolf said he put two out there, and did not know whatever happened to the other one.

Rolf said he ran a few ads, but word of mouth was how he got most of his repeat customers. He thought being mentioned in what Bob was doing and some of those early ads had helped business, but one of the other reasons Rolf said he quit running ads is because Bob started a canoe outfitting business called Canadian Border Outfitters in 1966, about seventeen miles north of Ely. He runs his business, and still does his writing and artwork, and has been active in the

---

[17] A copy of the letter from Bob Cary to Rolf is included in the Appendix.

wilderness preservation efforts. Rolf gave him advice and help when he started his business.

As we were getting ready to head out late that morning, the father and son Rolf outfitted the day before came walking up to the building. I overheard him and Rolf talking, and Rolf offering to refund part of what he had charged. Rolf was sorry the trip had not worked out, but he reminded me later that he thought it might not.

We jumped in the car and went down to the landing to collect the gear. Rolf thought he better get there and try to salvage as much food as possible. Unpacking, cleaning, and putting things away took time and put us behind schedule for the morning. I could tell Rolf was disappointed, not so much for the loss of revenue, but for the loss of the opportunity to see someone get the wilderness in them.

We worked on the cabins again in the afternoon, and at the dinner table Gail told me there was a group of girls and boys from Wisconsin staying at the campground. She talked to the leaders that afternoon, and mentioned that I might be interested in joining them around the campfire. They said they would be glad to have me, and she thought I might enjoy being around some people my age. They were camping over near the rapids and Gail said I could go over after dinner.

I took Waysha for a quick walk after dinner, something I did frequently. It is more accurate to say he walked me, though, because huskies are bred to pull. After we finished the walk, I headed over to the campground. I thought they would be a larger group and easy to find by listening for the noise because I could hear them long before I got there. The wilderness is so quiet and sound travels. Shrieks from boys and girls let me know this was a co-ed group.

As I entered the camp, the leader, a man in his twenties, came over and introduced himself. He made me feel welcome and introduced me to the group. It was a large group, about fifteen altogether, and most were kids about my age. The other leader, a female, was attending to one of the girls in a tent who seemed to be in pain and moaned from time to time. The female leader came out a couple of times to report to the other leader that the girl did not seem to be getting any better. Apparently, she was suffering from sharp abdominal pains that were getting worse.

On her third trip out, and as the darkness fell, she said she thought they ought to do something, but she was not sure what to do. I offered to get my employer who could take her to the Wilderness Witness Base.

"Would you please?" she said. "I don't think she can walk, and I don't know how we could get there."

With that, I started into the pitch black woods without a flashlight. It was light when I came over, and I never thought about a flashlight. I could not see where I was going, and there were no lights to be seen anywhere. I hoped I was walking straight and not in circles like you can do in the wilderness. Branches hit me in the face, and I kept praying that I would come to the road soon. I could not believe I had not thought to bring a flashlight.

I finally reached the road and recognized where I was. I was then able to start running to get Rolf. I saw the light on in the outfitting building, and hoped Rolf was there doing paperwork. He frequently did paperwork and returned phone calls after supper. Thankfully, he was.

"Rolf, we have an emergency over at the campground," I said, catching my breath. "A young girl is in great pain, and they wondered if you would take her down to the nurse at the Wilderness Witness Base."

"Sure, let's go," he said, as he grabbed his keys and headed out the door.

I filled him in on the way over, and Rolf always kept a flashlight in his car. We entered the campsite and could hear the girl now sobbing and obviously in great pain.

The female guide and two other girls helped get her into the car and they took off down the trail. Rolf was gone for about an hour, and when he returned, he told me that the nurse decided to take her down to the clinic in Grand Marais. She told Rolf she would call when she knew more.

A few days later, we got a call about the young girl who had been taken to the clinic. She had to have an emergency appendectomy because her appendix was about to rupture. The girl was resting fine, and they thanked us for our prompt willingness to help. Later I received a note[18] from the leader of the group personally thanking Rolf and me.

Ever since that incident, I have never ventured out into the wilderness without a flashlight. I had a real sense of panic for a few moments when I could not find my way, but I also realized how precious every moment can be in an emergency. Being far away from everything like we were, and farther if we had been on the trail, means you need to think ahead at all times. Thankfully, this event had a happy conclusion.

---

[18] A copy of the thank-you note is included in the Appendix.

CHAPTER 7 - MORE ABOUT THE GUNFLINT TRAIL

The summer continued with the usual schedule until one day around lunchtime when Rolf received a phone call from Sigurd Olson. Sigurd told him about a group called the Voyageurs who were coming to our area. The Voyageurs were a group of men who were reenacting the journeys of fur traders by traveling the routes used in the eighteenth and early nineteenth centuries. They started way up in Canada, north of Lake of the Woods, and came down to Lake Superior. It had been quite a news story in the northern areas, and the group was two weeks away from coming through our part of the old route. Their group consisted of four large birch bark canoes that were manned by a team of six paddlers for each canoe and the support team that was covering the event and making sure everything went well.

After Rolf finished talking and hung up the phone, he explained that Sigurd wanted to make a grand entrance from our location. He said the group was planning to camp near Clark Island when they passed through here, and they were going to have a cookout and celebration that night on Clark Island.

"Who is Sigurd Olson?" I asked.

Rolf told me that Sigurd was a well-known author in the area. He made many trips throughout the wilderness and wrote about them. He had a place in Ely, and he did a lot to protect the wilderness area, including speaking before Congress about his work for preservation. He was also a surveyor and teacher, and fought to have the wilderness left alone. The name Way of the Wilderness actually came from one of his books. Cal Rutstrum, another author, sold Rolf the property to start the outfitting business, and a quote from Sigurd's book was the inspiration for the name. Sigurd talked about the way of the wilderness being the way of the canoe, and Rolf liked the quote so much, he chose it as the name for his outfitting business. He said he had a real connection with both Cal and Sigurd.

The story goes that Cal Rutstum had a place over on Seagull Lake and owned the ground where the outfitting building is. He thought there ought to be an outfitting business there, so he asked Cliff Waters if he would be interested in starting a store. He also asked Rolf to set up an outfitting business. That was in 1956, so Rolf bought five new canoes and worked with Cliff to rent them the first year. Gail and Rolf kept their regular jobs that first year, then bought fifteen more canoes from Justine Kerfoot at Gunflint Lodge.

Figure 27: Gail and Rolf at their cabin on Seagull Lake

The second year is when they really got started and outfitted out of the back of Cliff's store and a small building Rolf built to hold paddles and other gear. Cal wanted Rolf to start the business and then buy the property and business from him after it was going. It was like buying your own business, but that is exactly what they did.

He and Gail had tried to buy Camp Windigo from Harry Brown, who owned an existing lodge over on Seagull Lake, but a few days before they were going to buy it, the government offered more money and that bid was accepted over theirs. They took down the buildings at Windigo and moved them to Poplar Lake. Rolf and Gail ended up where they were, expanded the business, and built the outfitting building after the government bought the land from Rolf that he had on Seagull Lake.

I asked him what happened to the cabin, and he said he bought it back from the government for five dollars when they bought the property on Seagull Lake. They told him to move it off the land over the winter or it would be burned. He said he took it apart and moved it over the ice by himself in his pickup truck. He not only moved his cabin, but took Cal's cabin apart and moved it that winter as well. Rolf's cabin eventually was sold, and the man who bought it put it back up on the Seagull River along the north shore of Seagull Lake, about a half mile south of the Seagull landing.[19] He told me Cliff got the best deal when he bought Peter Blomberg's house, which was the one Rolf helped build. Cliff got the building at auction for thirty-five dollars, took it down, and used the lumber to build the house he has.

I asked Rolf if he ever wanted to expand the business and grow bigger. He told me not really, that he liked the small operation. He said when he would occasionally stop in at Gunflint Lodge where they would have fifty kids in the water practicing capsizing drills. He said it was like a zoo. He preferred staying small and serving the families and smaller groups.

"I'll let Justine monkey with the large groups," he said.

I was impressed that Rolf knew most of the authors and other well-known people in the area.

"I think I know most of them," he said. "In addition to Sigurd, Cal, and Bob Cary, I know Ben Ferrier, who takes trips up to Hudson Bay. He wrote a book called, *God's River Country*[20]. Pipesmoke Gary, who used to come up here before

---

[19]Rolf still has Cal's cabin in pieces on his property at the End of the Trail.
[20]See Bibliography.

World War II, wrote a book called *Wither Away*.[21] Bill Magie is active in keeping the wilderness preserved.[22] Most of these guys have had a hand in preserving the wilderness, and all share a deep appreciation and love for the country.

I remember one time at Gunflint Lodge when Cal and Ben Ferrier were arguing about some portage. I sided with Ben because all he did was take trips every summer. Cal took a few trips, but he was more of a researcher. I even told Cal that a couple of his suggestions in his book were actually dangerous. He told me that the people reading his books lived in New York City apartments and read to escape their day-to-day lives. They were not really going to try these things about camping.

Figure 28: Rolf and Cal Rutstrum

They're all great guys who have done a lot for the wilderness, especially Sigurd. Out of all of them, I think I'm more like him. I just really want the wilderness preserved so people can enjoy it in the years to come. Sigurd's signed some

---

[21]This book is apparently out of print. No verifying information could be found to include in the Bibliography.
[22]David Olesen published a book in 1981 that includes stories of Bill Magie, called, *A Wonderful Country*. See Bibliography.

of his books for me and he has always thought of Saganaga Lake as one of his favorites. It will be good to see him."

"Gee, Rolf, you seem to know everyone up here," I said.

"Well, not everyone, but a lot of these guys who have a real love for the country, like I do. Most of those guys live over in the Ely area," Rolf said. "Gail knows these guys, too. When she came up here, she worked at the End of the Trail Lodge. That is where we actually met. When you stay up here, you get to know people."

We headed over to have lunch with Gail and the kids, and I was anxious to ask Gail about her story. I had plenty of time with Rolf, but not much with Gail, and I really wondered what brought her to the Northwoods.

Rolf mentioned to Gail that Sigurd called, and they chatted briefly about the Voyageur event. She also brought us up to date on the blueberry picking done that morning. Gail loved to pick blueberries and had several secret places she would go during the summer to pick. She picked strawberries in June and now, along with the blueberries, she was picking black raspberries. Taking all the kids was a job, but they usually had a nice haul.

HOW GAIL AND ROLF MET

After lunch was finished and the kids settled, I asked Gail and Rolf how they had met.

"Well, that's quite a story," Gail said."I was living in Fort Collins, Colorado, and attending college there. I had worked in several parks in Colorado, and had heard about the woods in Minnesota, and it sounded interesting. I answered an ad that Al Hedstrom, who owned End of the Trail Lodge, had placed in an employment office in Duluth. He had a standing order for summer help and waitresses, so I answered it and he wrote me back. He told me to get to Duluth, and take a bus from Duluth to Grand Marais, and to call him. He would come and pick me up. I remember that ride up the trail, wondering where I was going, and Al telling me not to get mixed up with any of the guides. I don't think Al thought I would last. I actually didn't meet Rolf, though, until the end of the summer."

"That was actually kind of funny," Rolf said. "Al had asked me to help him with a building, so on Labor Day, I closed up the cabin on Seagull and headed over. I got there about ten o'clock in the morning. Al had taken a group fishing and his wife Mary suggested that I wait for him to come back. She said she had a girl there working all summer from Colorado who had been wanting to go fishing, and hadn't had a chance to go yet. She said she was over at the other building hooking up a stove pipe, and that maybe I should see how she was doing. I had heard about Gail, but I hadn't met her because I was too busy and our paths had never crossed."

"I was over there all right," said Gail. "And as soon as Rolf walked in the building, that pipe came a-tumbling down. I was so embarrassed and wanted to get out of there, but Rolf introduced himself, and gave me a hand putting things back together and we began talking. When we went back to see Mary, she had fixed lunch and suggested that Rolf take me fishing. So he did."

"When was that?" I asked.

"That was in 1952," Gail said.

"I had been fishing all summer long, so it wasn't a new thing for me," Rolf said. "But it was for Gail."

"I took Rolf by surprise when I asked to run the motor. Right after we shoved off, I asked him if he would mind if I drove and he navigated. He was agreeable, so that's what we did. We caught our limit fishing over near Saganaga Falls, and then stopped and visited with the people at Canadian Customs who all knew Rolf. He introduced me around to everyone.

When we got back to the Lodge, we all had dinner, and we didn't really see each other much after that. Just a couple of social events, I think."

"I thought she would leave and I wouldn't see her again," Rolf said. "She talked about finishing school; and that she still had two years to go. We just said good-bye. I did get a Christmas card from her though."

"What happened next?" I asked.

"Well, I came back because Al offered me a better job with more pay," Gail said. "I drove back in my 1949 Nash sedan. My folks had traded it in for a new car, so I went down and bought it. Rolf had worked all winter building more things for Al, but was guiding for two other resorts. We saw each other more that summer, as he would come around after the day's work ended. I actually saw him more in the fall, as he was over at our lodge working on the buildings he had started in the winter. But after the season, I went to Duluth instead of home because a girlfriend I had met offered me a room and got me a job at Minnesota Power and Light. Rolf had a friend in Duluth who he came down to visit, and he called me when he was in town. That's when things began to get serious.

We got married February 13, 1954, and shortly thereafter is when the deal to buy Camp Windigo fell through. We were wondering how we could make a living just being resort hands, but Al wrote and invited us both to come work for him. I was going to get a better salary as head of the kitchen, and he made Rolf the head guide over thirteen guides. We saved a little, since room and board was free.

That winter Al headed to Florida, and Rolf had a friend in Taconite Harbor, which is on the shoreline on the way to Duluth, who offered him a job in the plant there. We stopped there and he hired Rolf and gave us a new trailer to stay in. Rolf added on to the trailer and I found a secretarial job. We ended up staying there two years because the work was steady."

"The second year we were there is when we started the outfitting business with Cal with the five canoes," Rolf said, continuing the story. "Then the next year is when we added the fifteen more and came up to run the business."

"That's quite a story, Gail, I said. "I didn't know you were that much of an outdoor kind of person. I guess I should have, given that you are here and married to Rolf. I also didn't realize you liked to fish."

"I love to fish. I just don't get to do it as much with the kids and all," she said, pointing to the west where the campground was as she continued talking about fishing. "I remember taking Sandy over to the Seagull River over by Campsite #13 when she was five. We were catching walleye left and right, and when I cut a branch off to string the fish on through their gills, we could hardly carry them home. I love the outdoor life."

"Is it hard to keep track of the kids up here?" I asked.

"More so in the summer when they're out of school. Sandy helps now that she's older. I do remember one day when Stanton, their special needs child, decided to go fishing and took off in a canoe by himself. We couldn't find him at first and then Rolf saw the canoe with Stanton in it offshore. Rolf took off in the boat and brought him back. Stanton had even remembered to bring a life jacket. We were scared at first, although we can laugh about it now."

With that, Rolf and I went back to our afternoon work and finished our plan for the day.

THE VOYAGEURS COME THROUGH

The day for the visit from Sigurd Olson and the Voyageurs finally arrived. It was early afternoon when Sigurd arrived at Rolf's building, and he looked just like the pictures Rolf had shown me. He was dressed in khaki pants and a matching shirt with a brown outdoors-man full-billed hat. He was fidgety that day, fumbling with his pipe and looking out the window often, hoping to see the people who were supposed to escort him out to Clark Island for his grand entrance.

He and Rolf visited quite awhile catching up on things. Luckily, we were not that busy, so I was able to listen in on the conversation. Apparently, the newspaper and outdoor magazine folks were supposed to escort him to the island and do a story on the event. They were late and Sigurd seemed quite annoyed.

Around 3:30 that afternoon, Rolf finally convinced Sigurd to let him run him out to Clark Island in his boat. Sigurd agreed, and the three of us took off for the celebration. It was quite an event. There were four huge birch bark canoes built to scale from the originals that were used by the fur traders of long ago. The island was abuzz with activity. Fires were burning and people were milling around. The actual campsite for the Voyageurs was in a different location, which provided more room on Clark Island for the festivities. Apparently, the group had arrived not long before we did, having paddled from Ely, some twenty-five miles away that morning. The festivities were already under way when we got there.

We had to find a place to pull in because the large canoes took up space and the boats of the visitors were moored and tied together to enable everyone to climb to shore. The cooks were getting ready with the fires to cook moose burgers for dinner, and the locals had brought bags of potato chips, potato salad, hot dogs, ice tea, and cookies for the dinner. Not quite the diet of the real voyageurs, but this was a special evening of celebration because it marked the last leg of their journey.

After a few minutes of surveying all the activity, one of the Voyageurs came up and started talking to me. I assumed he was one of the crew because several of the men wore a red kerchief for a bandana, and he was one of them. He was about five-feet four inches tall, with a dark tan, and muscular. When he started speaking, I could tell by his accent he was from Canada.

"Are you from around here?" he asked.

"Well, yes and no," I replied. "I'm working for a canoe outfitter this summer, but I live in Ohio."

"What do you think of all this?"

"Well, it seems pretty cool, but I don't know much about it," I said.

He told me this was a reenactment of the fur trading that followed these routes from way up in Canada down to Lake Superior. They started several weeks ago and were on the last leg to Lake Superior from here. He explained that much of the United States was mapped and settled on horseback and covered wagons because it was mostly land. Much of Canada was charted and mapped by canoe since so much of Canada, especially the area around here, was lakes, rivers, portages, and wilderness. You had to travel by canoe because traveling by any other means of transportation was often impossible. He said an enormous amount of the world's fresh water flows through Canada, and most of it was flowing through the route they had been following.

They started out of Ely at daybreak that morning, which was a distance of about twenty-five miles, and made it here by afternoon. He spoke with his hands while he talked, using a sweeping motion to allude to the whole country north into Canada.

"My goodness, you came from Ely since this morning?" I asked with amazement.

He said they moved fast on the water, and run the portages just like the real Voyageurs had done. The canoes they used were birch bark, fashioned after the ones made by the Ojibwa and Algonquin Indians. The pieces of bark had been carefully peeled from the tree and were laced together with spruce-root lacing and soaked in boiled pine sap to protect them. Later on, some of them made and used wooden canoes because they held up better in the rapids.

He said most of the guys in those days were small like him, because they took up less space in the canoe, allowing more room for the cargo. They did not eat as much as a bigger fellow either, so they were more suitable for this kind of work. The paddles were about fifty inches long, and a smaller man could handle them and paddle hard all day. They usually paddled fifteen hours a day, and sometimes as much as eighty miles in a full day. It took five or six guys to a handle a canoe because they were over twenty feet long and heavy from cargo. He told me they had six men in each of the canoes for the reenactment.

They mostly ran the portages, carrying as much as two-hundred pounds of stinky fur with a tumpline. The tumpline was wrapped around a bale of fur and placed across the forehead. The first bale rested on the bottom of the tumpline about hip high and behind you as you faced forward. With a bent back, a second and third bale was placed in the tumpline on top of the first. They brought furs along with their gear as a part of the reenactment to make it more authentic.

I said I was familiar with tumplines and asked what kind of food they ate.

In the early days, the voyageurs ate a small, pressed cake that had dried meat mixed with fruit, berries, and various kinds of animal fat. It was called pemmican. Sometimes they added flour and water and made it a soup dish. They would also make a paste of flour and water with dried peas and bacon added to it. They would catch fish or hunt game if they had time or were kept ashore by storms. That was always a treat.

When the voyageurs first started, there were conflicts with the Indians. Later, however, they became friendly with the Indians and traded with them. In turn, the Indians would help them navigate or fix their canoes. I found all this very interesting and was disappointed when he said he needed to go get ready for the canoe show.

I walked over to the fire pit and loaded a bun with a moose burger and all the fixings the locals provided. The burger was quite good, but had a little bit of a wild game taste. All the food was good, even the second burger I had.

By the time I finished eating, the Voyageurs were on the water and were performing amazing maneuvers with the canoes. First, they demonstrated the stroking speed they maintained while underway, and then did all kinds of stunts, passing their paddles among each other and between the canoes. They did a figure-eight stunt and after a couple of loops, they squared off with two canoes heading straight for each other. Just when they appeared to be on a collision course, they passed within two feet of each other, and the paddlers actually stroked in between each other as they passed. It was very entertaining and the spectators were thoroughly impressed. The show lasted quite awhile and I visited with several people around me.

I did not see much of Rolf and Sigurd. They always drew a crowd because of Sigurd's celebrity status in this area. Every time I looked for them, they had at least five or six people around them. I decided to join them and listen to the conversation. After a few minutes, Rolf moved toward me and told me he had been catching up a little with Tempest Powell.

Figure 29: L to R: Tempest, Marion and Ben Ferrier, Gail, and Irv Benson

Tempest was married to Irv Benson, and Rolf had known her for years. She was one of the Indians with whom he used to guide. She and Irv lived up on the eastern part of Saganaga Lake on an island. In fact, Rolf drove Irv to the government office when he wanted to buy the island. He said Irv told them he only wanted five acres, but the government official kept insisting that he buy the whole island, which was ten acres, and Irv kept insisting that he only wanted five acres. After a while, the government representative said he could not parcel the island and asked Irv if he was married. Irv said yes, so they put five acres in his name and five in Tempest's name, all for the same price. Tempest's mother was Mary Ottertail, the daughter of a chief from an Ojibwa band up on Lac La Croix.

Mary married Jack Powell and they lived up on Sagnagons. They had a log cabin up there and trapped, hunted, and generally lived off the land. They raised five children, and Tempest was the youngest.

Rolf said Tempest really knew certain areas of the country, but surprisingly, there were certain areas that she knew nothing about. She told him they only went into town (Ely) a couple of times a year, and they went a certain route every time. She had never been over the Silver Falls Portage until Rolf showed her where it was.

"You are kidding!" I said. "That's right next door to where they live!"

"I know," Rolf said.

He said they used to go a different route to Ely. She was in her thirties and guiding with Rolf before she had ever been down here on Seagull, Red Rock, and Alpine lakes. Of course, when she was younger, the Gunflint Trail had not really been finished out to Saganaga, so Ely was the largest town nearest the boundary waters. When you live up there like they did, you do not really have time to be tourists and explore. You are usually busy right where you are unless you have a resort like her brother, Frank Powell, who is married to Charlotte, and runs the Green Forest Lodge on the north shore of Saganaga. Even though Rolf showed her some new areas, she showed him areas on Sagnagons and around there that he did not know about, too, such as different portages.

She and Irv still live off the land there. Tempest had three daughters that Irv adopted when they got married, and they manage a sixty-mile trap line with seven cabins along it to work. According to Rolf, they have done pretty well selling furs to the Canadians. The beaver, mink, fisher, and other furs they harvest are pretty high-quality furs, and they sell well in England.

Tempest always had dogs, even though snowmobiles had been getting more popular. Irv always said he would only buy a snowmobile if the dogs could pull it.

"I remember a funny story about Tempest," Rolf said. "One December, I was working on the ice road to haul some tear-down cabins from Saganaga Lake over to the Wilderness

Canoe Base camp for Ham Muus. Ham had arranged for some volunteers to help over the Christmas holiday so I was pushing to clear a path on the ice to haul those cabins while the help was there. The ice was not quite solid in some places, and I was working on a section by the winter portage over near the narrows leading out to here, which is the main part of the lake. The winter portage is there because the narrows don't freeze up in the winter because the water moves through too rapidly. That's why it's marked on the map as the *winter portage*. You only need to use it in the winter.

Anyway, the back wheels of my truck broke through the ice and I was stuck. I chopped a hole in the ice about twenty feet in front of my truck and found a nice birch log to drop down in it. I winched the truck out by wrapping the winch cable around the log and pulling against it. It's a standard trick when you break through the ice. While I was working on it, here comes Tempest, snowshoeing across the lake with her dogs. She asked if I was catching any fish. I explained what had happened and we had a good laugh. Of course, that was big news for that day, and word got around the trail that the ice was still weak in places."

I asked Rolf if he had a connection to the Powells that run the resort on Saganaga. He said not really. He knew Frank and Charlotte, and Charlotte's sister, Betsy, who also worked out there, but they normally came into Chik-Wauk Resort when they came in, and he was at the End of the Trail Lodge, so he did not know them quite as well.

As the evening wore on, we got back in Rolf's boat and headed for home. Seeing this reenactment reminded Rolf of the artifacts he had found over the years. He thought some of them dated back to the actual voyageur time period. Over the years, Rolf found knives, shotgun shells with solid brass casings, cooking pots, and tin match containers made in England and France in the 1800s. He has found pipes and even a voyageur's belt.

Items made in England and France made their way down the St. Lawrence Seaway to the ports in the northern part of the state. Many artifacts could be found by digging down a little. There have been burns over the years so you have to get under the surface to find things. Some of the current portages are different from the original portages, but if you know where the old portages were located, you can find the original campsites and find artifacts easier because nobody is using those sites now.

Figure 30: Metal match holders left by the original voyageurs

As usual, Rolf kept the throttle wide open to get back. When we finally arrived, I hit the cot and fell sound asleep.

THE FINAL DAYS OF SUMMER

The final days with Rolf and his family were business as usual, however, I did have the chance to hear a few more new stories. I remember asking Rolf if he had any close calls or scary moments.

"Well, I have passed a bear on a portage before," he said. "That was kind of startling. I came up over a little hill and the bear was walking right toward me. He moved off the trail a little, and we kind of passed each other. Luckily, we didn't get tangled up. I also had a bear come around the outfitting building, and he left claw marks on the side of the building. I'm not sure what he was looking for.

I even remember one day when, out of the blue, Waysha started barking wildly and leaping up on the tree I kept him tied to. When I went back there to see what all the noise was, I saw a bear cub clinging to a tree branch about twenty feet in the air. I think what happened was that the mother ran him up the tree not seeing the dog. He was probably in his doghouse. That's how mothers part from their cubs, you know. They run them up a tree and leave them there, move on, and from then on the cub has to fend for itself. Anyway, I like to never got that dog away from that tree. He was pretty wound up.

I really haven't had many close calls because I try to be safe and not take any chances. I do remember a trip I took with two guys and their sons up through the Falls Chain toward Kawnipi. That was a little scary. It was two men and their fourteen-year old sons. For some reason, the boys were together in one canoe, and I had the dads with me in my canoe. The boys were kind of full of it and didn't listen as well as they should have. I probably should have split them up. Anyway, we made the Silver Falls Portage and Dead Man's Portage okay, and headed into the river that leads to the Falls Chain.

It was June, so the water was high and running pretty swiftly, and I paddled up to the boys and told them to stay back about a hundred yards until we unloaded at the portage. I told them that the current was stronger than they might think and if they didn't hang back until they could head right into the portage, they could get swept over the falls.

As I was unloading the last pack out of our canoe, I caught sight out of the corner of my eye that the other canoe was about ten feet away, moving right into the current to the falls. It's hard to see how fast you are moving unless you watch the shore as a gauge. They had no idea they were heading for the falls. My first thought was to yell at them, but I thought they might panic and skip a stroke or paddle the wrong direction. I caught their eye, waded waist-high into the water, told them to back paddle, and grabbed the stern of their canoe just in time.

The boys then realized how close they came to disaster and talked about it the rest of the trip. They actually became more cooperative after that and listened much better. Every falls up in that chain has claimed lives or caused serious injury for those who go over. They fall several feet with rushing water, and there is nothing but rocks all the way down. Canoes are buckled, gear and passengers are thrown, and the rushing water slams everything against the rocks. The Indians used to say that only fools run white water.

That trip seemed long because the kids kept using my ax to split logs, and hitting rocks with it. I was forever sharpening it. They wanted to help cook, and flipped more pancakes on the ground than in the pan. I was glad to be home after that trip. That was the closest I ever came to a tragedy on a trip.

You know, speaking of the Falls Chain, I remember a knife story up there. We had headed up through that same area one time and when we got up to Kawnipi to set up camp, I reached for my knife and it wasn't there. It was a Western style knife, and I had made a leather sheath for it to hang on my belt. I kept a pocketknife as well so I ended up using that the whole time. I couldn't figure where I may have lost that knife.

On the way back, I was carrying the canoe on the last portage out of the falls before Dead Man's Portage, and while I was looking down to check my footing in a muddy spot, I saw a shiny glimmer out of the corner of my eye. I set the canoe down and kicked around a little bit, and uncovered that knife. I was so surprised, because it could have been in the water, or someone could have picked it up. But none of that was true. I found it in the mud. I only saw about a half-inch of it glimmering, and only because I was looking down. That was a real find."

"You were awfully lucky to find that knife," I said.

"Another scary moment was when I was living up on Seagull in the cabin and I heard three gun shots ring out in rapid fire, striking my chimney pipe above the cabin," he said. "I almost came right out of my chair. It was early winter and I hadn't seen anyone for a couple of weeks, so I headed to the door to see what in the world was going on. Sure enough, it was Willard Waters, who owned an automatic carbine rifle. He was Cliff Waters' brother, and kind of a practical joker. He must have walked most of the day just to do that. He had a nice place over on Gull Lake and I had helped him build a hanger to house his pontoon airplane. I did some guiding for him, and he did some pilot work and drop-in business for parties who wanted to be dropped on a lake. He eventually sold his place and moved to Florida, and became a millionaire selling land in the Naples area in the fifties. He stayed the night; it was too late to go back with the ice being tricky at that time of year. Boy, did that raise my eyebrows!"

Moments like these with Rolf were rare. We continued to sit in the outfitting building while things were slow, talking back and forth. When I had moments like these, I generally kept listening because I was always fascinated to hear his stories.

"So Rolf, what are some of the weirdest things that have happened up here?" I asked. "Weirdest things, huh?" he said. "I remember a couple of nurses who came up here for a canoe trip and brought their mattresses with them. They were not going to sleep on the ground, so they packed those mattresses right into their canoe and took off. That was quite a sight to see those mattresses piled on the top of their gear. Luckily, it wasn't raining."

Rolf let out a laugh as he recalled the picture in his mind, then added that he guessed they made it out okay.

"We also had a guy come up here who brought a bunch of vegetable seeds with him and was going to live off the land for the summer," he said. "He went out to an island somewhere on Sag Lake. He didn't last very long. It's nothing but rock up here. There are very few areas with enough topsoil to grow vegetables so that didn't work out too well."

The only one who's been able to grow things successfully is Benny Ambrose. Benny lives on Ottertrack Lake just west of the Monument Portage. He was originally from Iowa, and came up here in the 1920s and worked a little with Russ Blankenburg. Rumor has it that he didn't get along well with his stepmother. When Benny left home, he cut down a live hornet's nest, opened his stepmother's bedroom window threw it in, and ran away. I think he was fourteen at the time he left.[23]

Benny and Russ did some logging and mineral prospecting, which is why he liked it up on Ottertrack. A lot of the lakes up there seem to have an emerald sheen to them and Benny thought there might be some good mineral veins up there.

He married and had two daughters and after ten years, his wife decided she wanted to live in town and so they went their own way. He did some guiding and trapping up there, but he had a huge vegetable garden. He would load up compost from around the area and bring it back to his place in bags in the canoe, and build up the dirt he needed to grow things like potatoes, carrots, onions, lettuce, and turnips. He even had some nice roses. When he went home to Iowa, he would load up on topsoil and bring that back, too. His garden is getting pretty big now, and seems to keep him well.

Benny's quite a character, you know. He used to paddle and portage to the landing here and then walk down the trail to Grand Marais to play baseball. He tried to be the tough

---

[23]Chik-Wauk Museum and Nature Center on the Gunflint Trail, Grand Marais, Minnesota has information about the history and the people of the Gunflint Trail, including Benny Ambrose.

guy. Sometimes in the winter, I would see him coming across the lake and when he got closer, he used to open his jacket and loosen his shirt to show how he could brave the cold. Always being the man's man. He used to head into town to vote, too. He is the only guy they grandfathered[24] to be able to keep living in the Wilderness area when the government changed the wilderness rules. Both he and Dorothy Molter down on Knife Lake have a special exemption. They can stay in the area until they pass on. Ben has a cabin up there he has been working on since 1929, I think, and I don't think it's done yet. I used to stop and see him when I went through there. Anyway, he's the only one that I think was successful at vegetable gardening other than maybe the Indians up on Sagnagons."[25]

"Who's Dorothy Molter?" I asked.

"So far Dorothy's been able to stay up on Knife Lake," he said. "Originally, she was a nurse from Chicago. She came up to work the Pines Resort in the thirties, and fell in love with the wilderness. She has her own cabin up there and has a real nice flower garden. She uses broken paddles as a fence around it to keep the animals out. People make it a point to drop off their broken paddles to help her enlarge her garden area.

She also makes root beer in the summer and sells it to people passing through. She's known for that, too.[26] I know Benny used to encourage her to bring in more soil so she

---

[24] A grandfather clause is a legal term used when an old rule continues to apply in special situations, while a new rule applies to the future.
[25] After Ben passed away, his buildings were removed, however, there is a plaque commemorating him mounted on the rock where he lived.
[26] Bob Cary wrote and illustrated a book about her called, *Root Beer Lady: The Story of Dorothy Molter* in 1993. See Bibliography.

could grow more. I don't think she ever did.[27] But anyway, that's two who are known for growing things up here.

I remember one other weird thing that happened. I had outfitted a guy with a canoe and all of sudden after a few days, he comes walking down the lane and up to the outfitting building.

"You probably have this happen a lot," he said. "But I lost your canoe."

"Happens a lot?" I said. "I've never had someone lose a canoe!

He explained that he got turned around on a portage and decided to walk back in by following the sounds of the other boats and motors. He said he couldn't get the canoe through the wilderness so he ditched it.

After talking to him for a while, I got a general idea of where he may have left it, so I took off to try and find it. I decided to cut zigzags through the woods area hoping I could cover more ground, and luckily, I saw a small patch of blue a little ways off. I had rented him a blue belt for a life preserver and when he ditched the canoe, he dropped it right side up and threw the preserver inside. Luckily, it had rained overnight, filled the canoe with water, and the preserver had floated to the top and was draped over the side of the canoe. That's what I spotted. Anyway, I retrieved the equipment and got it back. People are funny sometimes."

Just as we were finishing our conversation, we were interrupted by one of the kids who told us it was time for dinner. We wrapped things up and headed in. At dinner, Gail took a few moments to tell me how she enjoyed my being there and hoped that I had enjoyed my time with them. I was leaving the next day, and Gail told me she would miss me

---

[27]After Dorothy passed away, her cabin was moved and rebuilt in Ely, Minnesota, where it is now a museum.

being around. I told her I had had a wonderful time, learned a lot, and was particularly glad to get to know her and Rolf better.

After dinner, I began packing for home, a truly bittersweet moment. I was going to miss this clean air, rustic environment, and the wonderful moments with Rolf and Gail. Rolf paid for a Greyhound bus ticket for my return home, and I called my sister to go over the schedule. It would take over twenty-four hours to get home with all of the stops made along the way. The only souvenir I was taking home was an oval laminated picture of a deer in the woods on the slice of one-inch birch about nine inches in diameter. I still have that picture hanging in my shop.

Figure 31: Gail and children

The memories and everything I learned, however, were the real treasures I was taking with me. The extended time up north had inspired me and truly solidified the wilderness inside of me. The wilderness made a man out of my cousins and me, and I understood first-hand the depth of what Rolf's simple words really meant. It was fun watching Rolf work with other people who, like me, were coming there for adventure and to get away from the daily grind. True professionals, regardless of their field, are happier and much more successful than someone who is just doing a job. Rolf was a true professional in the broadest sense of the word, and it was a privilege to have spent this time with him and Gail.

In the morning, I said my goodbyes to Gail and the kids, and loaded my bag into Rolf's station wagon for the trip down the trail. As we rumbled along, Rolf and I chatted a little about the summer and I expressed how much I appreciated the opportunity to stay with him. We also talked about how much still needed to be done on the cabins we worked on and how the rest of the summer might play out for business.

When we arrived in Grand Marais, it was not much of a wait before the bus pulled in to pick up the four of us who were waiting. The familiar smell of diesel fuel and the hum of the bus idling was enough to remind me that I was back in civilization again.

Rolf and I shook hands and said our goodbyes and good lucks, and the bus pulled away heading toward Duluth. The thoughts I had about my time up north were interrupted only by the stops along the way to Duluth. People got on and some got off, and the change of buses in Chicago went smoothly. The last leg to Cleveland went quickly, and I slept most of the way. I wrote down notes from my summer when I was awake. I was not sure when I would ever get back to that area again, because I had overheard Aunt Nancy and Uncle Bud telling each other that they did not plan to return the following year. I know our lives and schedules would only get busier as we got older, and getting together again would be tough.

CHAPTER 8 - FAST FORWARD

Dick managed to get up north for his honeymoon in 1972, and Bob took a couple trips in the early seventies with a friend, but after that, none of us went back to the Boundary Waters until 1997. Rolf sold the business to Bud Darling in 1976. Bud was a CPA in Duluth with experience in administration from his days working at the YMCA. Rolf felt like their kids were missing out socially because they were so far away from Grand Marais, so he and Gail thought it would be better for them if they lived in town.

I reconnected with Rolf and Gail by telephone early in 1997 by finding their number in the phone book. Rolf answered and remembered exactly who I was, as did Gail. I was surprised since it had been thirty years since we last spoke. I told them that Bob and I were planning a canoe trip in the summer with our families, and we wondered if we could stop by and visit before we headed up the Gunflint Trail. Rolf said they would love to have us stop.

It was frustrating preparing for the trip because most of my gear consisted of what I had for state park camping like canvas tents, Coleman stoves, and iron skillets that only required being unloaded from a van. None of this gear would work on a canoe trip. Over the next few weeks, I found lightweight items that would work, and planned to rent Duluth packs and the canoes from Way of the Wilderness when we arrived. I was even able to find the original map I bought when my cousins and I took our five-day trip all those years ago. I wondered if it was still be accurate. It was about the only item I had that I knew was wilderness ready because it was waterproof.

Bob and I went back and forth several times as we prepared for the trip. We planned to stay in one of the new chalets Bud built when he had expanded the business. We would also use the tow service to American Point to the western section of Saganaga Lake. This was a new service offered at Way of the Wilderness. This would cut out about seven miles of paddling across the lake, and put us closer to the length of trip we thought our families could handle. From a business standpoint, I recognized that the chalets and the tow service would provide additional revenue streams for Way of the Wilderness, but I hoped the wilderness itself had not changed too much since I was there last.

Figure 32: Driving through Duluth

The trip we had planned was for four nights and five days and was of easy to moderate difficulty. We were launching from American Point and making a loop southwest through Ester, Hanson, and Eddy lakes, and then back up through Ogishkemuncie, Jasper, Alpine and Seagull lakes. It would be a good trip, and not too demanding for our first-timers.

My wife, Lynn, our boys Rob, Adam, and Will and I met Bob, his wife Carol, and their sons Dustin and Chad in Minneapolis. After a stop at Mall of America and a good night's rest, we headed north to the Gunflint Trail taking turns leading and following the way. I distinctly remember the elevation increase as we made our way toward Duluth, and that brought back memories from my first trip with Uncle Bud and Aunt Nancy.

Although Duluth had changed over the years with new freeways and buildings on the outskirts of the city, the downtown area was remarkably the same.

The lakefront development transformed the area into a tourist center and gathering place for concerts and local activities. I was thrilled to see familiar sights as we headed north along the shore of Lake Superior on Route 61. The little towns with their gift shops and tourist attractions were comfortably familiar, even though it had been so long ago. We passed the lakeside parks we visited as kids, and I remembered when we explored the creeks and streams that made their way down from the Boundary Waters to Lake Superior.

Figure 33: Route 61 Tunnel

Once we made our way through the Route 61 tunnel, I knew we were getting close to Grand Marais. The highway rises at the beginning of the city limits and once you reached the apex, the town can be seen spreading out along the waterfront. Rolf said that his street intersected with the highway and that we would see it on the left. We found it right away and turned up the hill to his house.

Although Rolf made it easy to find his house by using prominent house numbers, I immediately recognized the sixteen-foot Cadillac fishing boat with the Way of the Wilderness bear paw stencil on the bow. It was sitting on the trailer along the side of the driveway. I could hardly believe he still had the boat from thirty years ago, and a rush of memories flooded over me as I headed toward the door.

"Hello. How are you? Come in," Gail said as she answered the door with a big smile.

Rolf was right behind Gail, and motioned for us to come inside.

"It sure is good to see you," I said, as I reached out to shake his hand. "It sure is great to see you too, Gail."

Bob greeted Rolf and Gail, and soon we were making family introductions as Gail motioned us all to sit down. I noticed that the years had been kind to them. Other than the expected signs of age, they both looked almost the same as thirty years ago.

"Sit down everyone, and make yourself at home," Gail said. "I want to hear all about your plans."

We spent the next hour talking about what had happened over the last thirty years. Rolf and Gail told us about selling the outfitting business in 1976, the move to Grand Marais, and gave us updates on all their children. Their youngest, Sally, was just six months old when I worked there, and I anxiously listened to everything that happened to them over the years.

Gail worked at the U.S. Forestry office issuing permits, and Rolf had undergone heart bypass surgery. She was planning to write a book about the businesses along the Gunflint Trail. Unlike some of the books already written about the area, she wanted to have a special emphasis on the women who were involved in the area's history. It was not until after World War II, when the businesses and lodges were built and became known, that the area grew in popularity. It had remained relatively wild and unsettled until then. Since Rolf and Gail knew most of the locals and were a part of the history, Gail had a passion to capture the stories for posterity.

Rolf and Gail were very interested in our trip, and Gail was quizzing the two wives about their experiences and concerns about the trip as we got up to the present in our conversation.

"Do you know which way you're planning to go?" Rolf asked.

"I think so, Rolf," I said, as I began to unfold my thirty-year-old map and show him our proposed route. "Do you think this map will still be accurate? You know, I bought that from you on my first trip up here."

"I don't see why it wouldn't. The wilderness hasn't changed. The new ones have the campsites marked which is nice, but I still use my old maps. You can pick up a new one at Way of the Wilderness. Bud keeps all that stuff on hand."

I asked Rolf to look at our trip and tell us what we might expect to find and to point out areas of interest. He marked where there might be good fishing, and pointed out which lakes were known for what kind of fish. He also remarked that water in Ester Lake had an emerald green brilliance to it because of the mineral content. He told me to make note of that when we entered from the portage from Ottertrack Lake.

"Do you still have that can of dehydrated water?" I asked.

"Sure. Come with me," he said, as he stood up and motioned me to follow him to the garage.

I was hit with another flood of memories when we got to the garage. The can of dehydrated water, the picture of the Canadian Mountie and the Indian running the rapids, the Hamm's beer poster and sign, the map of Minnesota showing the Quetico wilderness as part of the United States were all there from the outfitting building where I worked for him thirty years earlier. It was comforting to see some of the special things I remembered as a young man. When Rolf handed me the can of dehydrated water, it was like stepping back in time.

Everyone else joined us as we were reminiscing about the items from the past. I even showed them the snowshoes Rolf kept in the outfitting building and the Duluth packs he rented and kept for personal use after he sold the business.

"Hey, Rolf, do you have any of that bug dope from World War II you used to sell?" I asked.

"The Insect Repellant? Sure, I still have some and I still use it. Would you like some?"

"Sure, I would!"

Rolf walked over to his car, reached into his glove box to retrieve something and handed a small bottle to me.

"Here you are. It still works. Just remember it only takes a little bit," he chided.

I was so thankful that we had taken the time to stop for a visit. As the conversation between Lynn, Carol and Gail was winding down, and we were preparing to leave, Gail reminded us to keep journals.

"Journaling is always a good idea so you can look back on your trips and remember little things that are easy to forget," she said. "Take time to write what you expect before you go, and then compare that with what actually happens. I always recommend that people keep a journal."

With that, we said our goodbyes, piled into the vans, and started the last part of the trip up the Gunflint Trail. As we drove along, I marveled over what had happened in the visit with Rolf and Gail. All those things I had neatly tucked away from my childhood were true. The stories I told over the years had been confirmed. Now I wondered what new memories the wilderness would stir.

I was amazed at how the ride up the trail looked the same as when I was a young man. The wilderness still came up to the sides of the road, although the road was now paved instead of graveled. The view of Grand Marais nestled along the vast expanse of Lake Superior was still spectacular as we headed north up the steep hill. Hedstrom Lumber Company was farther up the trail, and lodge signs popped out as we passed the side roads that led from the Gunflint Trail to where they were hidden in the wilderness. I rolled down the van windows and took a deep breath, inhaling the clean, refreshing air. I had already noticed that my sinuses were open, and my Ohio hay fever was subsiding.

I knew we were getting close when we made a wide turn overlooking Gunflint Lake. The water was a deep blue and the ripples and waves glistened in the sunlight, making a beautiful mirror of the wilderness behind it. Not much farther, we went through the S-turn by the big boulder and pulled into the parking lot at the end of the trail.

The right turn at the north end of the parking lot would lead us to the Way of the Wilderness outfitting building. When we made the turn, however, I did not see the building where I remembered it. It was not until we reached the end of the lane, and I looked to my left, that I saw the building sitting about one-hundred yards north and facing south. I could not help but wonder why it had been moved, and I wanted to know.

I knocked loudly on the door, but no one was there. We climbed up the hill to what used to be Cliff and Hilda's End of the Trail Store, and I found Bud Darling standing at the cash register.

"You must be Bud Darling," I said, introducing myself. "I used to work for Rolf Skrien."

"That's been a while ago," he said as he shook my hand.

I asked him if Cliff and Hilda were still around.

"No, I bought the building and business from them quite some time ago. However, it just so happens they're staying over at the End of the Trail campground in their trailer. You might want to take a moment to go say hello. They're all excited about watching the eagle's nest at the End of the Trail Campground with eaglets almost ready to hatch."

We spoke a little more about Bud buying Cliff's business, and about how the cafe was added as another revenue source. I noticed that the old adding machine Cliff and Hilda used when they owned the business and the original soda machine dispenser were displayed in the building. The flood of memories just kept on coming. We organized the permits and fishing licenses, and I gave Bud a list of the equipment that we would like to rent. Lynn and the boys were looking around the store at the post cards, maps, and things for sale.

Bud pointed out the chalets we rented for the night, and after we unloaded our gear, I took the family around to see the sights. We walked down by the Saganaga launch, then up to the End of the Trail Campground where I was able to find Cliff and Hilda's trailer. I knocked loudly on the trailer door.

Figure 34: Outfitting sign

"Hello, anybody home?"

"Sure, we're in here," Cliff said.

I introduced myself, and although they were polite about claiming to remember me, I am not sure they did. We had a brief conversation about their retirement and vacation and the eagle's nest they were watching. They looked and sounded the same as before, just older. I could not believe I was actually talking to them after all these years.

After our visit, we made our way over to the stream that runs into Gull Lake from Seagull Lake. The mosquitoes were thick, and we swatted as we walked. We arrived around sunset, and were pleasantly surprised with a gorgeous view of the orange sky peeking through the wilderness trees. The reflections of red, yellow, and orange glistening on the water made it seem almost three dimensional, and it was breathtaking.

With the sound of the water flowing and gurgling over the rocks in the background, I explained that the water was flowing north as we climbed the rocks on the nearby shore. We lingered for a while, taking in the beauty of the moment and the magnificent display God had provided for us to enjoy. As we started our walk back to the chalet, I knew there was something providential to my being there at this time to see Cliff and Hilda and to witness such a glorious display at sunset. Rolf had indeed gotten the wilderness in me, and now I hoped the wilderness would get in my family.

When we arrived back at the chalet, Bob and I began organizing things for our trip in the morning. We repacked our packs and got them ready for the tow out the next morning. I did not sleep well that night because of the excitement and the expectation of what the future held.

Figure 35: Sanderson family

We got up early, went to breakfast, and met at the launch point for our tow out to American Point. This was the point furthest west where motor boats were allowed to travel on Saganaga Lake. Saganaga and Seagull lakes allow motors in limited areas, and outside those lakes it is paddle only. Since past American Point is paddle only, that is the furthest point you can be towed.

It only took a few minutes to load the towboat with our gear, and we headed out across the opening that leads to the narrows. We noticed the cabins and other buildings as we made our way to the narrows. Although the government had removed or burned the structures in the wilderness area, the land at the trail's end had not been included in the preservation effort. The owners of the real estate next to the Boundary Waters had been grandfathered in, and were allowed to maintain businesses and cabins. During our visit, Rolf told me that he still owns five acres next to the Way of the Wilderness outfitting location. Just past the narrows is where the structures ended and the wilderness area began.

When we arrived at the narrows, the flood of memories returned of my first trip through in Uncle Bud and Aunt Nancy's fishing boat with the two canoes and Jim in tow. Dick was pointing out the rocks as Bud navigated the shallow water. Mark Darling did not ask us to help navigate because he made the trip almost daily.

Lake Saganaga was beautiful that morning. The sun was high in the sky, and the deep blue water sparkled in the sunshine. As we left the narrows, we could see the open water clearly. I could see Clark Island in the distance. I remembered our first year of camping there and the Voyageur celebration the second year I was there. It still looked the same.

There was a gentle breeze blowing from the west as we made our way to American Point. When we arrived, we unloaded the towboat and put our gear in the canoes. I had Adam and Will with me and decided to put Adam up front and have Will in the middle. Lynn and Rob teamed up in the other canoe. Dustin and Chad started out in one canoe, with Bob and Carol in the other, but this was a short-lived arrangement. Dustin and Chad kept bickering about the direction of the canoe, and eventually Bob teamed with Chad and put Dustin with Carol. This was the way it was for the rest of the trip.

We stopped for lunch at Swamp Portage and made the short five-rod portage, which is more of a nuisance than anything. It was a good practice run before the real portaging starts. We put back in and headed to Monument Portage. We encountered a little traffic on the portage because there were two parties already in progress. One of the groups included a family who had their grandfather with them. He told me that he had been coming up here for years, and his family wanted to bring him one more time. He could not move quickly, but he was glad to be there.

Figure 36: The narrows to Lake Saganaga

The short paddle to the next eighty-rod portage past the high bluffs on Ottertrack Lake brought us to the portage leading to Ester Lake. If we had gone farther south, we would have seen where Ben Ambrose lived. It was a challenging uphill climb from our direction, and I was glad to see the emerald green water sparkling through the trees at the end of the climb back down to water's edge. I remembered that Rolf had said about Ester Lake being known for its emerald color.

The high bluffs on Ester Lake reminded me of so many areas where we paddled as kids. Bluffs are common to the lakes and they majestically tower over them with their sheer faces plunging down into the bottom of the lakes. No two are the same.

We paddled through Ester Lake and through the reed narrows that connect to Lake Hanson. We found a nice campsite on Lake Hanson with a western exposure that allowed us to see the beautiful sunsets in the wilderness. We stayed there two nights, picking blueberries for the pancakes in the morning, fishing, and exploring. The mosquitoes were terrible at that time of year, and they drove us to our tents early in the evenings. Rangers visited us as we were breaking camp the second morning. They were checking permits, and we gave them the extra blueberries we had picked as a thank you for their commitment to the wilderness. Their job that day was to relocate and dig a new latrine, as the one on this site was getting full.

Figure 37: Lynn and Rob sailing

We made our way down the lake to the next several portages, the first one of which was infested with mosquitoes. I was the first to cross, and after dropping the canoe on the other side, ran back with the bug spray to spray everyone as they made their crossing. There were some fellows with mosquito head nets, and they helped the kids with their packs. Often people are helpful without being asked, and we appreciated their help. It was a beautiful portage, but we were glad to get out of there and back onto the water.

After a long day of paddling, sailing, and portaging in the hot sun, we arrived at Lake Ogishkemuncie. We set up camp, and all joined Lynn for a refreshing swim, something she had not been sure she would do.

After our swim, Rob and I headed back a ten-rod portage to do some fishing in a small lake next to Ogishkemuncie. Luck was on our side, and we landed a few northern pike, enough for a nice fish dinner that was a hit with everyone. There is nothing like tasty fresh fish caught in clean lake water, especially when prepared with my special fish fry recipe.[28] We enjoyed watching the birds gather, hover, and feast on the remains of the fish we left for them as we cleaned and filleted them. Rolf always said that the wilderness would take care of itself if you leave it alone.

We left the next morning to make the portages up to Alpine Lake, which would be our last night. It took us almost a day to paddle and portage, and that night we had the chance to finish the last of the food provisions. We spent a wonderful evening stargazing at the incredible sky. The Northern Lights were not visible that night, but we saw numerous shooting stars, and I pointed out the satellites that had become more visible in recent years. My family was quite impressed with how beautiful and rugged the wilderness was. We reviewed the highlights of the trip as we lay on the rocks and gazed at the sky. Bob and I longed to have a sip of apricot brandy like we did as kids. I made a mental note to remember the apricot brandy on all future trips for old time's sake.

We broke camp and got ready in the morning to make our last portage into Seagull Lake. The wind was behind us so we did some sailing, although it did not last long because Seagull Lake is so littered with islands. It took us some time to find the opening that leads to Seagull landing, and we understood why some people are easily turned around there.

Once we entered the narrows leading to the launch, we started a sprint to the finish as the launch came in sight. Bob and Chad were first in and won the bragging rights, and we

---

[28]See www.GettingTheWildernessInYou.com for the recipe.

started unpacking our gear. We returned the rented gear to Way of the Wilderness and transferred the personal gear to the vans.

We all enjoyed taking hot showers, which are always a highlight after getting off the trail, and we headed down the trail to Grand Marais. We stopped for a lunch of hamburgers and fries that hit the spot after being out, and then headed to see our friends again.

Rolf and Gail were expecting us and offered refreshments of coffee and soda when we arrived. We relived the trip and shared our stories with them, because I remembered how much Rolf and Gail enjoyed hearing the stories about the trips. They talked about the trips they took with their kids and their grandchildren.

Near the end of our visit, I asked Gail if it would be okay with her if we took a canoe trip the following year with Rolf. I was concerned that the bypass surgery might have weakened him, and I wanted to know what she thought of the idea. I was aware that he was in his mid-seventies as well. She thought it was a great idea, and the plans were then set for the trip. It would be our first trip with Rolf after all these years. Even though we had known him all these years, none of us had actually been on the trail with him. Bob, Dick, and I now looked forward to spending time with him on an actual canoe trip.

As we left Grand Marais, I thought about all that I had experienced in the last week. The memory of working there and finding so many familiar people and places that had been part of my life as a teenager was a comforting experience. I hoped my family enjoyed their moments in this special place. I knew it would probably not be as special to them as it had been to me because of the time in my life when I came here and the circumstances in my life as an orphan. Nonetheless, I asked my wife what she thought of the trip.

"Well, it was definitely hard, especially with a six-year-old and a nine-year-old," she said. "But it was spectacular in beauty. It's hard to describe beauty and rustic at the same time, but that's really how it was. I think it would have been easier as a first-timer to base camp rather than move in a loop. The mosquitoes were horrible, and I can't believe you didn't use the bug dope you got from Rolf."

I forgot it in the van, and it discolored the dashboard vinyl where it laid for the length of our trip.

"I absolutely loved the fresh fish, but not so much the dry food," she went on. "I thought the water from the lakes tasted good, and I enjoyed the refreshing swim, although I still can't believe I washed my hair in lake water. I'm really glad I came, and I enjoyed spending time with family, being Rob's canoe mate, and stargazing at night. You really get closer to people when you spend that much time together, and you really can't get away or go watch TV in another room. You have to work as a team or nobody succeeds. It was really a great experience. I would probably go again, but I'd prefer to base camp."

I smiled as I heard her words. I have never heard anyone express regret for spending time in this wilderness, and I thought that the wilderness was getting in her.

CHAPTER 9 - ONE MORE TIME

During the course of the year, Bob, Dick, and I communicated with each other, exchanging information on who had what and what we could use on the upcoming canoe trip. We also called Rolf to do as much planning as possible, so we would not have too much overlap of gear. Rolf and Bob had tents. Dick and I had cookware and other items, and Rolf offered his lightweight Grumman canoe that he had purchased in 1946, the same one the Indians had made fun of when he bought it, saying it would never last.

As the year went by and our plans to meet in September grew closer, a Northwest Airlines strike almost derailed the entire canoe trip. The strike finally settled in early September, just in time for our flights. We all planned to fly into Duluth, rent a car, and drive from there. I was leaving from Akron, Ohio, Bob was coming from Abilene, Texas, and Dick from Virginia Beach, Virginia. I had the good fortune of getting an early flight to Detroit with connections to Minneapolis and Duluth. I arrived in Duluth at 9:30 that morning, and rented a mid-sized car, thinking that it would be big enough for our drive up the North Shore of Lake Superior and the one-hour drive up the Gunflint Trail.

I was concerned about having enough room in the trunk for all the gear and about carrying Rolf's canoe, so I decided to check a few local garage sales to see if I might find a luggage rack to mount on top of the car. I had seen them occasionally at garage sales and as luck would have it, I found one for two dollars at the first garage sale on my trip from the airport. It was a gorgeous September day in Duluth with beautiful sunshine, and the temperature was a warm seventy-one degrees. Since Bob and Dick were not going to get in until four o'clock that afternoon, I had plenty of time to explore. I found a local tavern and had a nice salad for lunch. After lunch, I found a hardware store and bought a repair plate to fix my guitar because it had been damaged in flight. I knew Rolf would not mind if I used his drill to fix it before we left on our trip.

I headed over to the public park along the north shore of Lake Superior, and went for a long run along the water's edge, enjoying the beauty of the deep blue water of Lake Superior. It was a perfect day, ideal for a comfortable run. I even had time for a refreshing nap in the car with the windows down before I headed back to the airport where I was to meet Dick and Bob. Their plane arrived on time, and we were able to get their luggage and load everything quickly. We were soon on our way to Grand Marais.

We stopped at McDonald's on the way, and in his haste, Dick put his soft drink on the car roof. Without knowing it, we drove off with it there until we turned onto the main road. We felt like young boys again.

We made it to the motel, and checked into our room. We pulled our gear out of the car and headed to meet Rolf. He and Gail were glad to see us, and we spent the evening visiting, catching up, and talking about our gear and what we were going to do on the trip.

Our plan was to head south through Seagull, Alpine, and Jasper lakes and on to Ogishkemuncie. We were anxious to get down to Gabimichigami and Little Saganaga. The three of us had never been down in that section of the wilderness, and Rolf said he had not been down there for quite some time. He planned a few optional side trips for us in case we wanted to see where some of the old lodges had been. He thought we might find artifacts from the time when the lodges were active. I fixed my guitar while we were there, and we decided to meet early the next morning.

Bob, Dick, and I headed back to our motel room to finish packing for the trip, and go through the food and gear one last time. We organized everything in our packs and before you know it, it was time to turn out the lights for bed. As usual, sleep was hard to find that evening because we were so excited, and it felt like I only drifted off to sleep five minutes before hearing the alarm.

We arrived at Rolf's at 7:30 that morning and loaded his canoe on top of the car using the luggage racks I bought at the garage sale. As I feared, there was not enough room in the car trunk for all our gear, so we each had to hold a pack on our lap and tuck two packs under the canoe on the top of the car. Since it was a short ride, we were confident we could make it without being too uncomfortable.

Figure 38: L to R: Rolf, Bill, Dick & Bob

We arrived at the end of the trail later that morning around 9:00, had a little coffee, said hello to Bud and Mark Darling, and picked up our fishing licenses and the extra canoe we rented. We grabbed the Duluth packs and I noticed that Bud was selling some of the old Duluth packs from when Rolf owned the business. I stared at the table of old packs. They looked like the packs I helped stencil when I worked for Rolf. They were on sale for ten dollars each, so I bought two for old time's sake. I told Bud that we would need one less pack now since I was going to use one of the packs I just bought. I still own and use it today, even though it must be close to sixty years old. I gave the other one to my son Rob.

It rained all morning and kept raining while we packed our gear into the canoes in preparation for the launch at the Seagull Lake launch site. It was not a driving rain, but it was a nuisance, and strong enough to necessitate having on our rain gear. Rolf and I would be in one canoe, and Dick and Bob were in the other.

It was a little later when we pushed off, and we hoped to make it as far as Ogishkemuncie in a reasonable amount of time. The paddle across Seagull Lake was much different with Rolf on board. Instead of looking at the maps, we just followed the pointing paddle. Rolf would motion and point the direction we needed to go by lifting and pointing with his paddle. The only problem with paddle navigation is that you cannot easily learn navigation when you are alone. Rolf knew the lake like the back of his hand and that made it easier. During our paddle across, Rolf showed us where the lodges were and where his cabin was located with the two docks and three acres. He told us about Jack Miles buying Miles Island for $25,000 in the 1950s.

"They were nice people. In fact, I did some work for them. I always knew when they were pumping water, because I could smell the fumes from their gasoline pump over at my cabin," he said with a chuckle.

He pointed to where Cliff and Hilda and Cal Rutstrum used to live. He showed us the overhang in the rock where he would go to get out of bad weather. He showed us the island where Andy Mayo killed a bobcat with a boat oar, and casually mentioned that the cat was on display at the Chik-Wauk Lodge for many years. It was like having our own personal guide as we paddled along.

Rolf told us that he learned to winter fish as a boy, but learned more from the Indians who lived on Saganaga Lake when he first came there in the early fifties. He said you need two sticks after you cut a hole in the ice, one that was placed horizontally across the hole, and the second that was about fifteen inches long and placed vertically and onto which the line was tied. The vertical stick needed to have a Y crook that would hook onto the horizontal stick, when turned upside down, so it could not be pulled under the water. The vertical Y stick also needed to be long enough to suspend the line below the freeze depth of the water, so the fish would not cause the line to be cut on the sharp edge of the ice when they tugged. If the hole froze over night, which it was likely to do, you could chop through the ice without cutting the line. If you covered the hole with snow, it would slow the refreezing of the ice because it would act as insulation. It would also hide your hole from other fishermen or rangers who may be on patrol, looking for illegal fisherman.

Rolf explained that he would usually try to fish in the lake where the fish would most likely be, or would be likely to pass by. He said he used minnows or leeches as bait because they provided a little live action as well.

As we were leisurely paddling along, Rolf told me about the man who used rabbit meat as bait. One day he brought a rabbit to cut up for bait, and after he dropped his line into the water with the rabbit meat on it, he disappeared into the woods to relieve himself. There were several other people fishing, and while he was in the woods, the man closest to him reeled in his line (they were using fishing rods), and hooked the entire rabbit on the hook. He dropped it back in the water, so that the pole was bending from the weight when the man came out of the woods. When he saw his rod bending, he ran over and excitedly started reeling it in, only to see the rabbit carcass hanging on the end of his line. Rolf chuckled as he retold the story, saying that there is always room for a little fun even on the ice in the cold of winter. He added that winter fishing is much different from summer fishing.

There are not many people on the lakes in September, so we made good time with our paddle and portages. At the end of each portage, Rolf would carefully place some hard candy in sight on the top of a rock or log near the water's edge. He told us they were "portage pills" and would help make the portage more manageable. We got a good chuckle out of that, and indeed, they seemed to work. I make a point of packing portage pills for the trips I guide each year, and the parties always enjoy them. I even offer them to other parties we meet on our portages.

We found a good campsite on Ogishkemuncie and set up camp. We had time to do a little fishing around the coves and islands before we headed in to have our dinner for the evening. We had fresh chicken, carrots, and vegetables with us. The chicken had been frozen, but it thawed during our travel and tasted good over the open fire. I had several Coleman fire starters, and planned to use them to light the fire. Rolf brought along some handmade starters, and explained how he made them. He took a pressed paper egg carton and filled six of the ovals with paraffin wax and wood shavings. He broke one from the carton and handed it to me to try. I held my lighter to it and sure enough, it got the fire going and provided a nice long burn.

Since it was Sunday, I got out my guitar, and played gospel songs for a while. We sang until about 8:00, when the light drizzle that had been coming down became a more significant rain. We headed to the tents just ahead of a thunderstorm that began to light up the sky, when a large bolt of lightning struck not far from where we were camping. As beautiful as the wilderness can be, it can also be very intense. When a thunderstorm with high winds is raging anywhere in the lakes and wilderness, an individual seems so small and powerless. I remembered Rolf's words when I was younger. He said that the wilderness is absolute truth. It does not pretend, nor does it lie.

We broke camp the next morning after enjoying a breakfast of pancakes and morning devotions, then headed for two long portages. As we traveled, Rolf told more stories.

"Over there is where I shot a deer one year," he said, pointing to a section of the lake with his paddle. "The seaplane that had dropped me here was unable to take both me and the deer out at the same time. It was too much weight. So we decided he would take the deer first and then come back for me. You know, I thought just as he was pulling away, that if he didn't make it back, I would never get out of here."

Rolf chuckled as he continued.

"As it turned out, the seaplane finally returned an hour later to pick me up and we made it back. But I sure felt helpless for a few moments."

Rolf enjoyed telling his stories, and we enjoyed hearing them. He talked a lot about guiding, including one story about a couple on their honeymoon.

"I remember it vividly because the first night out it was raining like cats and dogs and we only had one tent. So I slept underneath the canoe to stay out of the weather," he said. "You can turn the canoe upside down and use two paddles wedged lengthwise under the seats and cross supports. Laying on these will keep you off the ground and out of the rain. It's not very comfortable, but it works."

I chuckled at his nonchalant recollection and piece of advice.

"Most of my guiding was smaller groups, especially groups out of Chicago," Rolf continued. "The first day or two was always a warm-up period, but after that people usually considered me part of the group, even the honeymooners."

We made the two long uphill portages and finally reached Gabimichigami, where we were going to set up our base camp. Rolf told us that Gabimichigami was the deepest lake in this area besides Lake Saganaga. He told us that lake depths are determined by instrumentation on airplanes, helicopters, and satellites, not by hand anymore the way they used to do it. He said that having ready, reliable data can help predict whether any major changes may be occurring.

"I think it actually is a good thing for monitoring the wilderness. Even fires can be watched better from the air than when you had to rely on ranger towers here and there," he said. "You know, they still require 4,000 feet of elevation for commercial flights over the Boundary Waters. It's just another way they've tried to preserve the wilderness."

We arrived at the campsite in the early afternoon and gobbled down a lunch of peanut butter and jelly on tortillas and apples. The wind started to build toward the end of our paddle, and the temperature dropped significantly. Rolf had

Figure 39: Rolf paddling in front of the canoe

a hammock that we hung between two trees and we all tried getting in and out of it without falling. Rolf was the only successful contestant.

We did not have much luck fishing, and spent most of our time paddling to keep from being blown into the shorelines. It was blowing so hard that we had to hang a tarp to break the wind so we could light a fire that evening. The dinner of chicken fajitas and songs on the guitar had a warming effect, but the wind chill made it feel like it was about twenty degrees, and when the rain started again we headed to the tents. Rolf continued to tell stories and mentioned the Indians who showed him where the burial stones were located. He told us about Chief Blackstone whose tribe died off one winter from a strange fever.

"They lived back in the bay off Kawnipi on the Canadian side, north of Sagnagons," he said. "They named a lake after the chief, Blackstone Lake. It was a sad thing. Without medical attention, everyone got infected and they all died."

The wind blew all night out of the north and had a bone-chilling effect. I was glad the tents were warm and dry. We got up on the third day at 6:30, and after a breakfast of scrambled eggs with hash browns, we bundled up with our hats and gloves, packed our lunches, and headed to Little Sag and Elm Lake, where we wanted to do some fishing.

Bob caught a Northern pike on Rattle Lake, and Dick caught a couple bass. It took us a while to find the little-used portage to Elm Lake, but we eventually found it, and made our way through with some difficulty. It was somewhat easier, however, with only the fishing tackle and a daypack. There were downed trees, evidence that not much traffic passed through that particular spot, but the lake was beautiful when we reached the end of the portage. Close to the end nearest the portage was a campsite area elevated to about twenty-five feet.

We decided to have our lunch there, relaxing and enjoying the beauty of this little lake. The clouds had broken apart, and the sun was peeking through to expose the beauty of the blue sky on this gorgeous fall day. The temperature had been rising steadily, so it was much more comfortable than the morning had been. When I closed my eyes, I could almost imagine a seaplane flying overhead and dropping someone on this little lake for an extended stay. What a great place to spend some wilderness summer days.

After lunch, we continued fishing from our canoes. We wanted to work our way around the shorelines, and talked about pushing up into a small, unnamed lake connected to Elm Lake. We were not having much luck fishing along the shore, so we turned toward the unnamed lake. We had to push hard through the seaweeds and reeds because there was no established portage. Once we got through we made our first casts, and the pike were hitting right and left. I caught ten in an hour, and released all but one. Rolf brought in several, and Dick caught a thirty-seven-inch pike that won the honor of occupying our dinner pot. We each contributed a dollar at the start of the trip for the winner of the biggest fish, so Dick had a few extra dollars in his pocket.

I was surprised we did so well. It seemed like a nonstop rattle of fish banging in the bottom of the aluminum canoes as we landed them, so we took the hooks out and released them. Rolf explained that this little lake would provide plenty of food for the fish and, because it was later in the year and the water temperature was cooling, they had migrated back into here. He also explained that this was a good breeding area for Northern Pike because of the weeds. We had a great time fishing in that little lake.

Figure 40: L to R: Bill, Bob & Rolf

It was not long before we headed back to base camp. Dragging and carrying fish on a portage is a good workout. Tired and yet thrilled, Dick cleaned the fish, and we had a delicious fish dinner that evening. We then spent time talking about the day and watching a few stars that peeked through the intermittent clouds. Rolf commented that pushing up into lakes that are rarely traveled was one of his favorite things to do.

The wind subsided, but a misty rain began to develop around bedtime, and the air temperature dropped down in the high thirties, meaning that it would not take much to see snow. Then the winds calmed down, and the sky cleared, giving us a beautiful starry night to view once again. We went to bed late that night, even though we knew we had a big day coming.

Our plan was to make it halfway back, so after another breakfast of pancakes, we got on the water early, hoping to complete the six portages and paddle by late afternoon. We had lunch on Lake Jasper at the Jasper Lake Portage, and unfortunately, the wind was in our faces made our return paddle formidable. Luckily, the sunny sky had a warming effect, and the air temperature improved as the day continued.

We finally made it to the campsite on Alpine Lake across from the portage that goes into Seagull Lake. This was the same campsite where Bob and I stayed on the last night of our trip the previous year with our families. We arrived at 4:30 in the afternoon, and we walked out on the flat rock that protrudes thirty or so feet into the water and stretched out for a while before making camp. It was one of those special wilderness moments. The warmth of the bright sun was absorbed by our dark wool and flannel clothing, and the soft cool breeze brushed across our faces. It was a perfect time for a refreshing nap after a long day on the water.

After our afternoon siesta, we set up camp, did some fishing, and enjoyed a dinner of lasagna, blueberry cobbler, and chocolate mocha pudding. During dinner, Rolf showed us how to feed the Canadian Gray Jays. He put a piece of bread on his fork and held it up with one hand out to his side. After a few minutes, one of the birds swooped down and grabbed the bread. As Dick loaded his fork with bread to try his luck with the birds, Rolf remarked that he had even seen them take bacon out of a frying pan.

Figure 41: Canadian Gray Jay looking for food

We sat around the campfire after feeding the birds while I played my guitar, and sang some more songs, told stories, and hit the sack early around nine o'clock that night. The last hard laugh we had was listening to Bob root through his entire pack for about five minutes. He had the loudest, crinkliest plastic liner and it sounded like he was holding a microphone next to it for amplification. He rooted and rooted for his pen, and finally found it after five minutes. He claimed he wanted to make some last-minute notes in his log and he could not find his pen. We all offered our extra pens, but Bob seemed to prefer the annoyance of the crinkling and kept searching.

The following day we woke bright and early at 6:30, and after a filling breakfast of hash browns and the rest of the fresh eggs, we were on the water to make the portage that leads to Seagull Lake. Luckily, the wind was at our backs and it did not take as long to travel across Seagull as it had the first day. Rolf told more stories about the old resorts, and showed us the rock where the eagles had been nesting for years.

We arrived at the Seagull launch around 10:30 that morning, unloaded the canoes, and loaded the car. We returned the rented items from Way of the Wilderness, said good-bye to Bud and Mark, and were back in the car an hour later. After dropping Rolf and his gear at his house, we said goodbye and thank you to each other, and promised to share pictures as soon as they were developed. Dick, Bob, and I headed to the Dairy Queen for lunch, and then on to Duluth where we were staying the night before our departing flights in the morning.

We were all tired and looked forward to hot showers. When I come back from a trip like this, I am always amazed to think that just a few short hours ago I was standing in the middle of the wilderness, miles and miles north, traveling the waterways by paddle.

After our hot showers, we decided to go to Grandma's Bar and Grill and have a satisfying dish of Minnesota wild rice with chicken teriyaki and mushrooms smothered in mozzarella cheese. It was a great end to our trip, and we spent time talking about our adventures and all the wonderful and fun things we heard and saw.

We turned the lights out before 9:30 that evening, and were up at 4:30 a.m. so Bob could leave to catch his flight. Dick and I woke up again about an hour later, packed our gear, dropped off the rental car, and made it to the airport in time to catch our flight to Minneapolis. We said goodbye as he headed to Washington, D.C., and I headed to Cleveland. The drone of the airplane engines caused me to drift in and out sleep, and my dreams were filled with all the wonderful things that occurred during my time in the wilderness.

I thought about how I had encountered the beauty and mystery of the wilderness as a young man, and how I had been able to take my family to meet the wilderness. I thought about being able to take a trip with the man who had introduced me to this wonderful area of the world. It was then that I knew that I had come full circle in my life. The wilderness was truly in me, and I could think of nothing better than to share it with others.

EPILOGUE

After our trip with Rolf, I took my boys Rob, Adam, and Will on that same route a couple years after our trip with Rolf. Dick, Bob, and I went back again, and I have been taking groups annually through a museum I founded in 2005. I have also been able to take my first grandson to the Boundary waters, and he loves to go. Plans for the other family members are underway. I always find it interesting to take people who are not family members and watch them get the wilderness in them. I now know how Rolf felt all those years.

Figure 42: L to R: Will, Adam, Bill, and Rob

In 2008, Dick, Bob, my friend, Denny, and I planned and completed the Hunter's Island trip, also known as the Diamond Trip. It is called that because the one-hundred and forty-mile loop forms the shape of a diamond on the map. That was a trip to remember with favorite moments like capsizing the fully loaded canoe only three days out. Every trip has special memories, and deserves a detailed writing that may happen in the future.

I have been blessed to have the simple enjoyment of sharing the wilderness with lifelong companions like Dick and Bob. I have been privileged to take strangers to the wilderness on our annual treks, and my life has been immensely broadened by these experiences. Sharing the wilderness with my family is profoundly special. Nothing was more thrilling than completing the loop with Lynn and the boys, and then to make the trip with just my boys a few years later. We still talk about these trips. My middle son, Adam, spent a summer working for Canadian Outfitters, the company originally started by Bob Cary around the time I worked for Rolf. Adam still uses his Duluth pack for hauling stuff around even when he is not on the trail.

The groups I take from the museum each year usually have never heard of this wilderness jewel, and are intrigued to make the trip because of my stories. Many of them have dreamed of adventures as boys and girls and longed for the time and place to experience this kind of journey. The thought of pristine surroundings, adventure, nature, and life outdoors seems to tug at the very soul of most of us. It has been exciting to see people's lives changed after being there. Our trips always include a stop to see Rolf, and he still shares stories each visit.

I am always confounded when I spend time with Rolf because I learn something else I never knew about the wilderness and the history of the area. The stories in this book are just a small sample of Rolf's profound knowledge of the area.

Figure 43: Moose and calf

He lives in Grand Marias, and keeps track of what happens there in the wilderness, by moccasin telegraph, I suppose. He spends his winters in Arizona and as much time with his children and grandchildren as he can. He still camps with his family, all of whom routinely make their way back north to the Boundary Waters, and will always be connected to it.

Gail passed away in 2002 before completing her book about the women and businesses of the Gunflint Trail. The committee she assembled completed the work in her honor, and called it, *A Taste of the Gunflint*. I always like to recommend journaling for the folks who take the annual trip with me because Gail was so fond of encouraging it. I also find many ways to share the wisdom and knowledge I learned from Rolf and others in my years of going up to the Boundary Waters. Like Rolf, I always enjoy hearing about the personal testimonies when these participants return. Several of them have gone more than once, and even plan to return.

I am convinced that there is great wisdom in Rolf's desire to get the wilderness in people. I finally understand the meaning of what he meant when he said that man cannot make the wilderness, but the wilderness can make the man. I have witnessed it in my own life and in the lives of those around me who have spent time in this special place.

I close my story where I started it, and I acknowledge that all of us experience things in our lives that change us. Sometimes these experiences are out of our control. Sometimes they seem random and we do not realize their importance until we get older. But we are never alone.

God created us to be relational, and He designed relationships to help us through life. Rolf has spent his life learning about and sharing his knowledge of the wilderness, and he touched many generations beyond his own. He made his love for the wilderness his life's work and used his influence in meaningful ways. I trust you have enjoyed our time together, and I hope to see you someday in the wilderness. We never know what a difference we make in someone's life. Rolf always said, "When you paddle into the wilderness, look for God. He's all over the place."

Pass the baton or maybe pass the paddle!

BIBLIOGRAPHY

Cary, Bob, *Root Beer Lady: The Story of Dorothy Molter*. Minneapolis, Minnesota: University of Minnesota Press, 1993.

Cary, Bob, *Tales from Jackpine Bob*. Duluth, Minnesota: Pfeifer-Hamilton Publishers, 1996.

Ferrier, Ben and Marion Ferrier, *God's River Country*. Englewood Cliffs, New Jersey: Prentice-Hall Press, Inc., 1956.

Henricksson, John, *Gunflint: The Trail, the People, the Stories*. Cambridge, Minnesota: Adventure Publications, 2003.

Hunt, F. Keith, *Look Over Your Shoulder: A Companion to Tough Men, Tough Boats*. Big Lake, Minnesota: Caira Press, 2004.

Muus, Ham (Herman A.), *Wilderness Witness the Founding Years, 1956-1966: Perspective on the Formation of Wilderness Canoe Base*. Digital Orange, 2006.

Nute, Grace Lee, *The Voyageur*. St. Paul, Minnesota: The Minnesota Historical Society Press, 1987.

Olesen, David, and Bill Magie, *A Wonderful Country: The Quetico-Superior Stories of Bill Magie*. Ely, Minnesota: Raven Productions, Inc., 2005.

Olson, Sigurd F., *The Singing Wilderness*. New York City, New York: Alfred A. Knopf, Inc., 1956.

Olson, Sigurd F., *Listening Point*. New York City, New York: Alfred A. Knopf, Inc., 1958.

Olson, Sigurd F., *The Lonely Land*. New York City, New York: Alfred A. Knopf, Inc., 1961.

Olson, Sigurd F., *Runes of the North*. New York City, New York: Alfred A. Knopf, Inc., 1963.

Proescholdt, Kevin, Rip Rapson, and Miron L. Heinselman, *Troubled Waters: The Fight for the Boundary Waters Canoe Area Wilderness*. St. Cloud, Minnesota: North Star Press of St. Cloud, Inc., 1995.

Renner, Jeff, *Lightning Strikes: Staying Safe Under Stormy Skies*. Seattle, Washington: Mountaineers Books, 2002.

Rutstrum, Calvin, *The New Way of the Wilderness*. Minneapolis, Minnesota: University of Minnesota Press, 1958.

Rutstrum, Calvin, *The Wilderness Life*. Minneapolis, Minnesota: University of Minnesota Press, 1975.

Webster, Janna, *Ki-osh-kons: People, Places, and Stories of Seagull Lake.* Janna Webster, 1997.

Women of the Gunflint Trail, *A Taste of the Gunflint Trail: Stories & Recipes from the Lodges as Shared by the Women of the Gunflint Trail.* Cambridge, Minnesota: Adventure Publications, March 2005.

APPENDIX

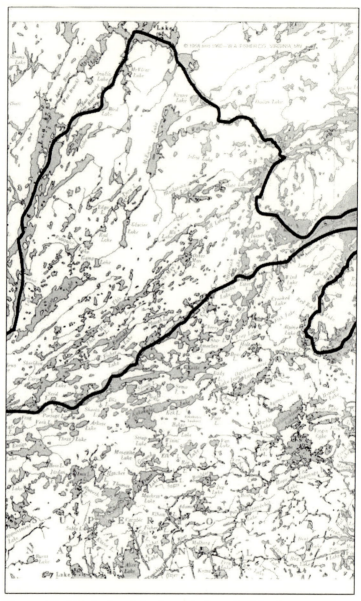

Figure A3: This is the eight-day trip described in Chapter 5.

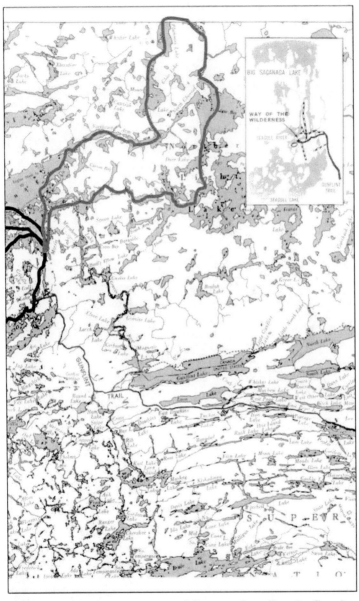

Figure A4: The loop at the top of the page describes the five-day trip from Chapter 3.

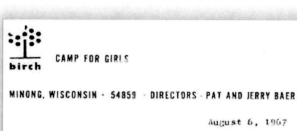

Figure A5: This is the thank-you note mentioned in Chapter 6.

New Canoe Outfitters Base To Open at Sea Gull

Rolf and Gail Skrien have leased a piece of land on Sea Gull lake next to the Cliff Waters place and will open a canoe outfitting business named "The Way of the Wilderness." The Skriens have an option to buy.

Mr. Skrien has been a guide on Sea Gull and Saganaga; since 1946 working for various resorts. He has lived on Sea Gull for three years.

He was born at Ulen, Minnesota, near Moorhead, and he attended school at Morris. He took three years of college work at Morris. Football fans will recall Dave Skrien of the University of Minnesota who was a power at fullback in 1948, 49, and '50. He is Rolf's younger brother. Dave was captain of the team in 1950. There are two other brothers in the family. His father is president of the Morris State Bank.

Mrs. Skrien is from Colorado. She attended Colorado State College at Fort Collins. The Skriens have one daughter, Sandra, aged 5½ months

Figure A6: Newspaper article from 1956

Vocational Biographies

Canoe Outfitter

Related Occupations
Resort Manager
Canoe Guide
Recreation Equipment Salesperson

Dedicated to the Preservation of the Wilderness

Canoe Outfitter Helps Vacationers Get Back to Nature

"I am always thrilled to see a group come back from a canoe trip—bubbling with excitement and telling of the things they saw and did," says Rolf Skrien. "Like a teacher, I feel I have contributed to their enjoyment of learning about nature and the great outdoors."

Rolf and his wife Gail, canoe outfitters, are owners of The Way of the Wilderness in the northeastern tip of Minnesota. Equipping visitors to the area with everything they need for a pleasurable, scenic canoe trip is their responsibility.

A few years ago the "back to nature" movement was a relatively new concept. People were waking up to the fact that there was more to life than concrete buildings, traffic jams, and polluted air.

These were the people who began searching through their dusty attics and basements for their old "boy scout" tents, their gas lanterns and their never-failing cooking stoves. The boom to get close to nature was on.

Many of these campers are discovering the joys of camping. They pack all necessary food, clothing, cooking utensils, shelter, and any other essentials in a canoe and paddle through areas that would be otherwise inaccessible.

And, unlike many kinds of camping, once the paddle dips in the water and the trip begins, the canoeists will not be able to return for forgotten equipment without beginning their trip all over again. Thus, Rolf's job and the supplies he selects for his canoeists are perhaps the most important factors of a successful canoe trip.

Learning to Work with Others

Attending Morris High School in Morris, Minn., Rolf's favorite classes were science, biology, and music. Extracurricular activities, sports, music, drama, publications, and numerous social events occupied much of his after-school time.

He also managed to work part time as a bellhop and in a dry cleaning shop after school. Rolf kept busy during his summer vacations working as a farmhand and driving for a veterinarian.

Figure A7: *Vocational Biographies'* out-of-print article featuring Rolf and Way of the Wilderness

After high school graduation, Rolf attended the West Central Agricultural School in Morris. Enrolled in a business course of study, he says he particularly enjoyed his shop and mechanical arts classes taken during his two years there.

Then Rolf moved to California and studied aircraft inspection at the San Diego Vocational School.

"Getting along with people and working well with others are probably the most important traits my formal education taught me," Rolf says. "Learning about mechanical structures and how to work with my hands have also been helpful in my career as a canoe outfitter."

High school students considering a career as a canoe outfitter should understand mechanical operations and be able to work well with others, Rolf advises.

"In addition, I suggest they take science, math, business, and bookkeeping classes to prepare them for operating a business like mine. Speech classes and participation in a wide variety of social activities are experiences that will really help prepare the potential canoe outfitter for working with all types of people," he says.

Summer work experience with a canoe outfitter, as a canoe guide, or even work at a resort is also excellent preparation for this type of career, Rolf adds.

Rolf worked at Solar Aircraft as an aircraft inspector before his stint in the military. Then he returned to Northern Minnesota to work as a resort guide.

During the next seven years, he was employed as a guide for canoe and fishing expeditions and learned about the resort business firsthand from others employed in the growing recreational field.

Rolf made many friends in this work. One of the most colorful was an old trapper who left quite an impression on him. The trapper helped implant some of the basic feelings about nature that Rolf now lives with daily: the camaraderie with the breathtaking beauty of the great north country, the respect and love for the raw wilderness, and the desire to keep it the way it is.

"I am completely dedicated to the preservation of the wilderness," Rolf says.

Totally Natural Setting

Located in a totally natural setting, The Way of the Wilderness is 60 miles from the nearest town.

"To get to this, the world's greatest canoe country, you travel to Duluth in northeastern Minnesota and take the North Shore Drive along Lake Superior to the Gunflint Trail," Rolf says. "Then you take the Old Gunflint Trail to the endmost point, and you've reached the best canoe country in the world."

Once the vacationers reach this area, Rolf is ready to help them prepare for their vacation in the natural environment.

"I average between 80 and 100 hours of working time each week from May to October, the heavy canoe season," Rolf says.

A typical day finds him renting canoes and equipment to about 20 people. Since he can supply both full and partial outfits, it is necessary for Rolf to help each party pick out equipment and gear, the amounts and kinds of food, the kinds of cooking equipment, the rain gear and tents, the type of canoe, paddles, and safety devices that each group of canoeists will need.

Then Rolf will help them route the kind of trip they want whether it will be for the best look at the scenery, for observing wildlife, for fishing, or for a more rugged expedition covering many miles and including portages across land.

The desired length of stay in the wilderness also influences the gear and the routes Rolf will recommend. Whether an overnight stay or an expedition lasting a couple of weeks, Rolf can help plan an individualized journey and a getaway from technology and urban problems.

Figure A8: *Vocational Biographies*, page 2

The outfitters might also sell bait and fishing licenses and supply additional information such as Canadian customs regulations.

Some outfitters also run lodges where people can eat and sleep before and after their canoe trip. Some outfitters cater to special groups such as college students or church groups.

At least one outfitter offers to fly his patrons in and out leaving them to camp and canoe as long as they want. One will even deliver the food to the landing site and have trained personnel to help the vacationers set up camp.

The possibilities seem to be as numerous as the outfitters, Rolf points out.

It is becoming a competitive business where the most successful canoe outfitter is the one who can offer the most for the least amount of money. Still both the American and Canadian governments are making the kinds of laws that will help people like Rolf Skrien keep this vast area natural and unspoiled. For him it's not only fulfillment of a goal; it's good business.

Preparing and Repairing

Rolf has some help during the busiest parts of the summer. All of the food and the right amounts of food for each group of canoeists must be packed in watertight containers but take the least space and weight possible. The maps needed to navigate in the wilderness must be in waterproof plastic.

It is necessary to continually clean and repair or service the equipment, canoes, boats, and motors as they return so they will be ready for the next group. Sleeping bag liners must be changed and laundered, and tents aired and cleaned. Rolf repairs his own boat motors. The Skriens also maintain a campsite for the convenience of their canoeists.

For Rolf, a quiet-spoken, good-natured outdoorsman, the necessary correspondence, advertising, and book work are the least liked duties of his business. He explains correspondence must be answered to give people information about his outfitting station and the equipment and services available.

The Skriens put out attractive brochures which could lure anyone to this spot they are so enthusiastic about. They also send out a concise list of all their available equipment and supplies, and round it off with good advice on what to bring yourself.

Rolf also indicates that it is necessary to advertise in national sports magazines and in the local papers. And, like any small businessman, he has book work to keep up-to-date.

"I think there is a great future in a career as a canoe outfitter or any outdoor recreation activity," Rolf says. "Anyone considering this type of a career must like to work with people and enjoy the outdoors to make a success of it.

"For a long time I was the only canoe outfitter up here," he says. "Now there are 23."

The off-season, during the late fall and winter months, is a time to clean and put away the equipment preparing it for the next season. Rolf takes inventory of equipment to be repaired or new equipment to be purchased. Some ordering, advertising, and correspondence is also done during the winter.

"The off-season is a time to catch up on anything that needs doing," Rolf says. "I also do some construction work for other people during this time. One outfitter I know teaches during the winter."

Rolf earns between $10,000 and $15,000 per year as a canoe outfitter. Being self-employed, however, means he receives no fringe benefits other than the ones he supplies himself.

Holidays and vacations are taken when he feels he can take some time away from work. He must supply his own hospitalization, accident, and life insurance as well as provide for his retirement.

Canoe Country is Why

Rolf and Gail, who has a background in resort work, met on a fishing trip. Although Gail is from Colorado, they both think Minnesota is one of the best states in America, and

Figure A9: *Vocational Biographies*, page 3

"This Canoe Country is why," Rolf says. They do not intend to move or change their occupation.

The Skriens have five children, Sandra, 16, Stuart and Stanton, 14, Susan, 10, and Sally, 7. The Skriens move from the canoe outfitting site into the town of Grand Marais for the winter so the children can go to school. Their winter activities take them outdoors, too.

"Winter or summer, I enjoy any activity out-of-doors," Rolf says. During the winter the Skriens especially like cross-country skiing.

Likewise, Gail loves outdoor activities of all kinds and reading. Together they enjoy canoeing, cross-country skiing, and traveling. Because they take care of other people's vacations in the summer, the Skriens' vacationing is done during the winter. Vacation travel takes them to Gail's home and friends in Colorado, and to the west and south.

Rolf is an officer in the Lutheran Church in Grand Marais, and he is active in recreation associations. Rolf likes to read, his favorites being travel and outdoor publications. Most of their friends are those they've met in connection with the canoe world and other people who enjoy skiing and outdoor activities with them

SUMMARY

More and more people are going back to nature on their vacations, and with this trend, many are discovering the joys of canoeing. The canoe outfitter is playing an important role in making this experience an enjoyable, educational, and safe one. He provides the proper gear and equipment plus good, correct information learned through study of the area and experience in traveling and routing safe scenic passageways.

Rolf and Gail Skrien are usually successful as canoe outfitters and owners of The Way of the Wilderness in northeastern Minnesota. Probably one element of their success is their intense love and concern for the raw wilderness and their desire to maintain the balance of nature. Rolf has been a student of this north country for years, and he and his family intend to stay in this area and in this business.

JOB FACTS
Salary Range:
 $9,000 to $15,000 per year

Educational Requirements:
 On-the-job training; work experience as a guide or at a resort helpful

Personal Characteristics:
 The ability to become familiar with the objects, materials, and services involved; drive and initiative; and the ability to work well with others

FOR MORE INFORMATION
The subjects of these biographies are not able to answer personal inquiries. For more information, please write to the following organizations:
National Recreation and Park Association
1601 North Kent St.
Arlington, VA 22209

American Association for Health,
 Physical Education & Recreation
1201—16th St. NW
Washington, DC 20036

Figure A10: *Vocational Biographies*, page 4

Figure A11: Hamm's Beer article from 1960s that featured the Hamm's bear and Rolf.

Figure A12: Hamm's Beer bear article, page 2

Figure A13: Early 1960's newspaper article featuring Rolf at his desk

Rolfe Skrien
Way of the Wilderness Outfitters
Grand Marais, Minn.

Dear Rolfe:

    Will be at Al Hedstroms July 6. Figure to jump off from McFarland Lake the morning of July 8. Will cut down thru Saganaga during the day.

    Would appreciate a cache of 4 cans of outboard gas, mixed with oil, at the Ottertrack side of the Monument portage. Figure 4 gallons will get me down to Basswood Lodge where I can get gassed up again.

    The reason for having it on the west side of the portage is to save time...since _ ..` .. ... & ..... that won't be busted the next week.

    I'll stop over and see you when I get into Al's. Will spend a couple of days canoing around...want to run the Granite river one day to get the lay of it.

    Hope this finds you and the family healthy and having a busy summer.

Best regards,
Bob Cary

P.S.: Publicity on this trip will be carried in the Chicago Daily News, Prairie Sportsman Magazine and the Spectator, Joliet, Ill. newspaper.

Figure A14: Bob Cary letter to Rolf referred to in Chapter 6

Figure A15: Second letter from Bob Cary to Rolf referred to in Chapter 6

Figure A16: Bill's original planning lists for the eight-day trip

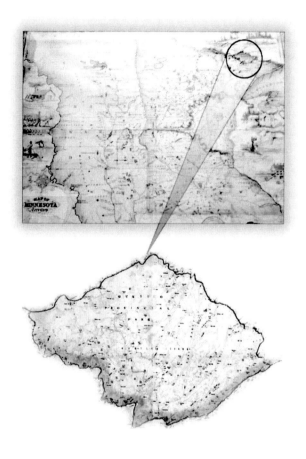

Figure A18: Map of Minnesota Territory mentioned in Chapter 2 showing the Quetico

WAY OF THE WILDERNESS CANOE OUTFITTERS
ROLF & GAIL SKRIEN — GRAND MARAIS, MINN.
Phone 388-7512

SUPPLIES FOR 2 PEOPLE FOR 7 DAYS

STAPLES	BREAKFASTS	DINNERS
1 lb. Shortening	3 lbs. bacon	Fresh Meat
1 lb. butter	1 doz. eggs	1 can each:
2 lbs. sugar	2 lbs. pancake flour	wieners
½ lb. salt	1 bottle Mapleine	dried beef
1 can pepper	1 lb. oatmeal	hamburgers
4 loaves bread	2 lbs. dried fruit	chicken fricassee
1 lb. Bisquick	1 instant fruit juice	beef stew
1 lb. fish fry meal	1 pkg. dried toast	mixed vegetables
16 tea bags		baked beans
1 lb. coffee	**LUNCHES**	corn
½ lb. inst. choc. drink	1 lb. cheese	green beans
3 qts. powdered milk	1 lb. sausage	peas
1 dehy. cream	1 jar peanut butter	tomatoes
3 pkgs. dehy. soups	1 jar jam	pears
CAMP SUPPLIES	1 can lunch meat	fruit cocktail
1 bar soap	1 can chili	peaches
1 dish cloth	6 pkgs. instant soft drink	2 pkgs. inst. potatoes
1 dish towel	1 box Rye Krisp	1 box inst. rice
2 candles	3 lbs. cookies	1 Kraft dinner
1 box waterproof matches	1 lb. raisins	½ lb. onions
1 box scouring pads	candy bars	3 lbs. potatoes
1 roll toilet paper		3 pkgs. inst. puddings
1 roll paper toweling		
1 roll aluminum foil		
1 bug bomb		

We adjust this basic list to the number in the party and the length or type of trip. Provisions are generous and are carefully selected from dried, concentrated or pre-cooked items as well as fresh or canned goods. Special lines of camp foods are also available in individual packages or complete kits.

Food is packed with regard to waterproofing and convenience in plastic and cloth bags or special cans.

Figure A19: Sample supply list for canoe trip

ORDERING INFORMATION

For more information about ordering
Getting the Wilderness in You,
please e-mail or visit our websites.

INFO@GetttingTheWildernessInYou.com
www.GetttingTheWildernessInYou.com
www.**AkronFossils**.com

ABOUT THE AUTHOR

William Sanderson owns Independence Financial Group, and has been in practice for over thirty-five years. He is a Chartered Financial Consultant and Chartered Life Underwriter, has a B.S. in Education from Miami University, and a Master's Degree from Malone University. He is a former member of Malone University's Graduate School Advisory Board, and has served as an adjunct professor there.

Bill's first canoe trip to the Boundary Waters Canoe Area Wilderness region of Minnesota was in 1966 with his cousins, where he met Rolf Skrien. The second trip was in 1967, when he then stayed on and worked for Rolf and Gail Skrien in their canoe outfitting business, Way of the Wilderness. Since then, he has returned many times, and annually takes groups from the Akron Fossils and Science Center, an organization in Akron, Ohio that he founded in 2005.

CPSIA information can be obtained at www.ICGtesting.com
Printed in the USA
LVOW041006040612

284554LV00001B/9/P